EPA/600/R-13/214A | September 2013 | www.epa.go/ncea

United States
Environmental Protection
Agency

I0488145

# Next Generation Risk Assessment:

Incorporation of Recent Advances in Molecular, Computational, and Systems Biology

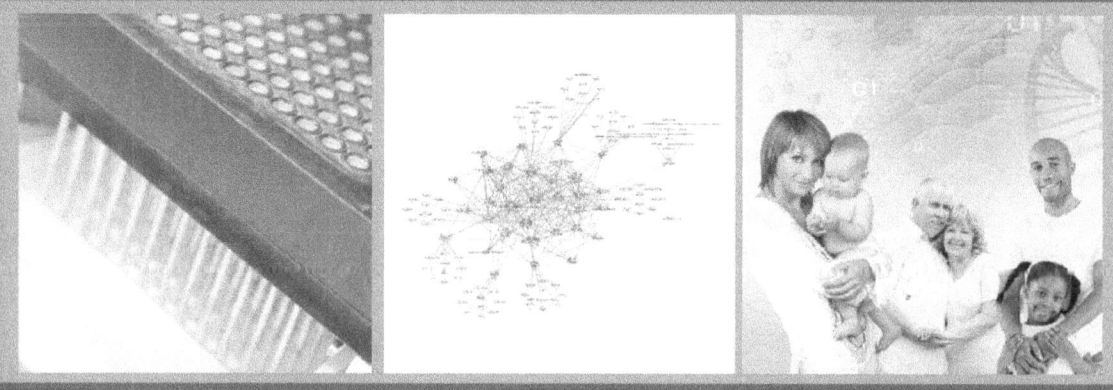

# External Review Draft

National Center for Environmental Assessment
Office of Research and Development, Washington, DC 20460

 United States
Environmental Protection
Agency

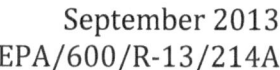
September 2013
EPA/600/R-13/214A

# Next Generation Risk Assessment:

# Incorporation of Recent Advances in Molecular, Computational, and Systems Biology

[External Review Draft]

National Center for Environmental Assessment

Office of Research and Development

U.S. Environmental Protection Agency

Washington, DC 20460

# Disclaimer

This document is an external review draft. This information is distributed solely for the purpose of pre-dissemination peer review under applicable information quality guidelines. It has not been formally disseminated by EPA. It does not represent and should not be construed to represent any Agency determination or policy. It is being circulated for review of its technical accuracy and science policy implications. Mention of trade names or commercial products does not constitute endorsement or recommendation for use.

*This document is a draft for review purposes only and does not constitute Agency policy. DRAFT – Do Not Cite or Quote.*

September 2013          ii

# Contents

---

*This document is a draft for review purposes only and does not constitute Agency policy. DRAFT – Do Not Cite or Quote.*

*This document is a draft for review purposes only and does not constitute Agency policy. DRAFT – Do Not Cite or Quote.*

September 2013                                        iv

# Acronyms and Abbreviations

| Acronym Abbreviation | Stands For |
| --- | --- |
| $AC_{50}$ | concentration at 50% of maximum activity |
| AER | activity-to-exposure ratio |
| AhR | aryl hydrocarbon receptor |
| AML | acute myeloid leukemia |
| AOP | adverse outcome pathway |
| B[a]P | benzo[a]pyrene |
| BMD | benchmark dose |
| CDC | Centers for Disease Control and Prevention |
| $C_{ss}$ | concentration, steady state (in blood) |
| CTD | Comparative Toxicogenomic Database |
| DEG | differentially expressed gene |
| DNA | deoxyribonucleic acid |
| EPA | U.S. Environmental Protection Agency |
| EWAS | environment-wide association studies |
| GEO | Gene Expression Omnibus |
| GWAS | genome-wide association studies |
| HCS | high-content screening |
| HPT | hypothalamus-pituitary-thyroid |
| HT | high-throughput |
| HTS | high-throughput screening |
| HTVMD | high-throughput virtual molecular docking |
| $IC_{50}$ | concentration producing a 50% inhibition of response |
| $IC_{10}$ | concentration producing a 10% inhibition of response |
| IVIVE | *in vitro* to *in vivo* extrapolation |
| KB | Knowledgebase |
| LEC | lowest effective concentration |
| MIE | molecular initiating event |
| MOA | mode of action |
| mRNA | messenger ribonucleic acid |
| NCEA | National Center for Environmental Assessment (EPA) |
| NexGen | Next Generation Risk Assessment |
| NHANES | National Health and Nutrition Examination Survey |
| NRC | National Research Council |

*This document is a draft for review purposes only and does not constitute Agency policy. DRAFT – Do Not Cite or Quote.*

September 2013                v

| Acronym Abbreviation | Stands For |
| --- | --- |
| NTP | National Toxicology Program |
| OECD | Organization of Economic Co-operation and Development |
| PAH | polycyclic aromatic hydrocarbon |
| PK | pharmacokinetic |
| POD | point of departure |
| ppb | part per billion |
| ppm | part per million |
| QSAR | quantitative structure-activity relationship |
| RNA | ribonucleic acid |
| ROS | reactive oxygen species |
| SNP | single nucleotide polymorphism |
| SOAR | Systematic Omics Analysis Review |
| Tox21 | Toxicology in the 21st Century |
| VARIMED | VARiants Informing MEDicine |
| VT | virtual tissue |

# Acknowledgments

This document reflects contributions of many individuals whom we gratefully acknowledge. Without the generous contributions and collaborations of these scientists this report would not have been possible. Appendix A lists technical papers related to this report and provide an more complete list of contributors. We would also like to acknowledge Ms. Rebecca Clark, Dr. Kenneth Olden, and Ms. Debra Walsh for their continued support of this project.

## Report Authors and Managing Editors

Ila Cote, Lyle Burgoon, Robert DeWoskin—EPA Office of Research and Development

## Authors and Contributors

### *Executive Summary*
Elaine Cohen Hubal, Rob DeWoskin, Lyle Burgoon, Ila Cote

### Introduction, Preparation for Prototype Development and Consideration of Decision Context
Ila Cote, Paul Anastas, Stan Barone, Linda Birnbaum, Becki Clark, Kathleen Deener, David Dix, Stephen Edwards, and Peter Preuss

### A Framework
Daniel Krewski, Margit Westphal, Greg Paoli, Maxine Croteau, Mustafa Al-Zoughool, Mel Andersen, Weihsueh Chiu, Lyle Burgoon, and Ila Cote

### Science Community and Stakeholder Engagement
Kim Osborn, Gerald Poje, Ron White

### Recurring Issues in Risk Assessment
Daniel Krewski, Melvin Andersen, Kim Boekelheide, Frederic Dois, Lyle Burgoon, Weihsueh Chiu, Michael DeVito, Hisham El-Masri, Lynn Flowers, Michael Goldsmith, Derek Knight, Thomas Knudsen, William Lefew, Greg Paoli, Edward Perkins, Ivan Rusyn, Cecilia Tan, Linda Teuschler, Russell Thomas, Maurice Whelan, Timothy Zacharewski, Lauren Zeise, and Ila Cote

### *The Prototypes*

### Major Scope Assessments- Benzene-Induced Leukemia
Ila Cote, Reuben Thomas, Alan Hubbard, Cliona McHale, Luoping Zhang, Stephen Rappaport, Qing Lan, Nathaniel Rothman, Jennifer Jinot, Babasaheb Sonawane, Martyn Smith and Kathryn Guyton

### Major Scope Assessments - Ozone-induced Lung Inflammation and Injury
Robert Devlin, Kelly Duncan, James Crooks, David Miller, Lyle Burgoon, Michael Schmitt, Stephen Edwards, Shaun McCullough, and David Diaz-Sanchez

### Major Scope Assessments Benzo[a]pyrene, Polycyclic Aromatic Hydrocarbons, and Cancer
Lyle Burgoon and Emma McConnell

**Risk Assessment Implications across the Tier 3 Prototypes**
Lyle Burgoon, Rob DeWoskin and Ila Cote

**Tier 2: Limited Scope Assessments - Knowledge Mining – Diabetes/Obesity**
Lyle Burgoon, Shannon Bell, Chirag Patel, Kristine Thayer, Stephen Edwards

**Tier 2: Limited Scope Assessments Short-Term In Vivo Models – Alternative Species**
Edward Perkins, Gerald Ankley, Stephanie Padilla, Dan Peterson, Daniel Villeneuve

**Tier 2: Limited Scope Assessments Short-Term In Vivo Models – Mammalian Species**
Michael DeVito, Jason Lambert, Scott Wesselkamper, Russell Thomas

**Tier 1 QSAR and High-Throughput Virtual Molecular Docking (HTVMD)**
Rob DeWoskin, Nina Wang, Jay Zhao, Scott Wesselkamper, Jason Lambert, Dan Petersen, Lyle Burgoon

**Tier 1: High-Throughput and High-Content Assays, Toxicokinetics, High-Throughput Exposure Estimation, Virtual Tissue (VT) Modeling, Thyroid Example**
Kevin Crofton and Richard Judson

*Advanced Approaches to Issues in Risk Assessment*

**Human Variability including Genomic Variability**
Lauren Zeise, Frederic Bois, Weihsueh Chiu, Ila Cote, Dale Hattis, Ivan Rusyn, Kathryn Guyton, and Lyle Burgoon

**Early Life Exposures**
Ila Cote, adapted from a paper by Boekelheide et al. 2012

**Mixtures and Nonchemical Stressors**
Timothy Zacharewski, Ila Cote, Linda Teuschler, and Lyle Burgoon

**Inter-Species Extrapolation**
Lyle Burgoon, Ila Cote, and Edward Perkins

**Low Dose-Response Modeling**
Weihsueh Chiu, Dan Krewski and Lyle Burgoon

*Conclusions*

**Lessons Learned from Developing the Prototypes**
Ila Cote, Rob DeWoskin, Lynn Flowers, John Vandenberg, Douglas-Crawford, Brown, Lyle Burgoon

**Challenges and Next Steps**
Elaine Cohn Hubel, Tina Bahadori, Ila Cote

Special Thanks to:

ICF International: Kim Osborn, Heather Dantzker, Jessica Wignall, William Mendez, Bruce Fowler, Codi Sharp, Deshira Wallace, and Pam Ross

Agency Partners who provided staff, data, ideas, and reviews

U.S. Army Corps of Engineers: Edward Perkins and Anita Meyer

U.S. Department of Defense, Office of the Secretary of Defense: Robert Boyd

California EPA: George Alexeeff, Martha Sandy, Lauren Zeise

Centers for Disease Control, National Center for Environmental Health, and the Agency for Toxic
    Substances and Disease Registry: Chris Portier (retired), Thomas Sinks, and Bruce
    Fowler (retired)

European Chemicals Agency: Derek Knight

European Joint Research Commission: Maurice Whelan

FDA's National Center Toxicological Research: Donna Mendrick and William Slikker

Health Canada: Carol Yauk

National Institutes of Health, National Center for Advancing Translational Science:
    Menghang Xia

NIH National Institute of Environmental Health Sciences:
    Linda Birnbaum, Scott Auerbach, John Balbus, Michael DeVito, Elizabeth Maull, Kristine Thayer,
    and Ray Tice

National Institute for Occupational Safety and Health: Christine Sofge, Paul Schulte, Ainsley Weston

EPA

Office of Research and Development:
    Michael Broder, David Bussard, Vincent Cogliano, Rory Conolly Dan Costa, Sally Darney,
    Elizabeth Erwin, Susan Euling. Annette Gatchett, Gary Hatch, Annie Jarabek, Robert Kavlock,
    Channa Keshava, Monica Linnenbrink, Matt Martin, Connie Meacham, Shaun McCullough.
    David Miller, Kenneth Olden, David Reif, James Samet, Rita Schoney, Imran Shah, Deborah Segal,
    Woodrow Setzer, Michael Slimak, Rong-Lin Wang, Debra Walsh, John Wambaugh, Paul White

Office of Air and Radiation: Souad Benromdhane, Bryan Hubbell, Kelly Rimer, Carl Mazza,
    Dierdre Murphy, Susan Stone and Lydia Wegman (retired)

• Office of Chemical Safety and Pollution Prevention: Stan Barone, Vicki Dellarco, Steven Knott,
    Mary Manibusan, Jennifer McLain, Jeff Morris, Anita Pease, Laura Parsons, and Jennifer Seed

Office of Superfund and Emergency Response: Michele Burgess, Rebecca Clark, Helen Dawson,
    Stiven Foster, and Kathleen Raffaele

Office of Water: Cynthia Dougherty, Elizabeth Doyle, and Elizabeth Southerland

Office of Children's Health Protection: Michael Firestone, Brenda Foos

Office of Environmental Justice: Charles Lee

Regional Liaisons: Carole Braverman, Bruce Duncan

Other Contributors

Ken Ramos, University of Louisville

Peter McClure, Heather Carlson-Lynch, and Julie Stickney, SRC

Catherine Blake, University of Illinois

William Pennie and Karen Leach, Pfizer

# Executive Summary

1   The Next Generation (NexGen) of Risk Assessment program was initiated in 2010 as a multiyear,
2   multi-organization effort to consider new molecular, computational, and systems biology
3   approaches for use in risk assessments. The goal is to enable faster, less expensive, and more robust
4   assessments for chemicals and other stressors that might adversely affect public health and the
5   environment. Although this report is focused on human disease, the approaches described here are
6   equally applicable to environmental risks. Specific aims of this initial phase of the NexGen program
7   are to (1) demonstrate proof-of-concept that the data and methods from recent advances in biology
8   can better inform risk assessment; (2) identify which of the information resources and practices are
9   most useful for particular purposes (value of information); (3) articulate decision considerations
10  for use of new types of data and methods to inform risk assessment; and (4) identify important data
11  gaps.

12  To achieve the above, prototypes or case studies were designed to (1) implement the
13  recommendations from workshops and experts on approaches to identifying and evaluating the
14  available data in molecular, computational, and systems biology for use in risk assessment;
15  (2) provide risk assessors, risk managers, and the general public with clear examples
16  demonstrating how new data and advanced methods might support specific types of risk
17  assessments; and (3) elicit interest, discussion, and participation from stakeholders to further
18  improve risk assessments.

19  The assessment prototypes are broadly categorized into three groups based on the assessment's
20  "fitness for intended use" given the decision context. Primary drivers of the decision context are the
21  number of chemicals that must be addressed and the confidence needed in the scientific data to
22  support a specific type of decision. The three categories or tiers have been defined as follows:
23  Tier 3—major scope decision-making (considerable data indicating high hazard or widespread
24  exposures); Tier 2—limited decision-making (limited exposure potential or limited hazard
25  potential or data); and Tier 1—prioritization and screening (very little or no traditional data for
26  chemicals known to be in commerce). Although the prototypes were designed for illustrative
27  purposes to address these three types of decision context, the supporting data and methods can be
28  deployed across all categories as available and as needed, and are arrayed as a continuum of
29  approaches. Ideally, multiple data streams are brought to bear on consideration of potential risks.

30  The prototypes illustrate types of data and methods that are likely to be used in the near future, but
31  are not intended to be exhaustive reviews. The primary intent of the first set of chemicals (Tier 3
32  prototypes) is to verify if and how new data types and approaches could be used to inform risk
33  assessment by comparison to robust traditional "known" risk, thus verifying new approaches. The
34  intent of the Tier 2 prototypes is to (1) explore new types of computational analyses and short-
35  duration *in vivo* bioassays that are relatively uncommon in risk assessment but appear very
36  promising for the near future; and (2) develop an assessment approach well suited to limited scope
37  risk management decisions. In this case, limited generally means regional to local exposure
38  potential, or limited hazard potential, or limited data to conduct more detailed assessments. The
39  Tier 2 efforts fall between Tier 3 and Tier 1 in terms of resources required and amount of
40  uncertainty in the assessment results. The Tier 1 prototypes explore entirely high-throughput
41  approaches that could be applied to thousands or tens of thousands of chemicals, are the least
42  resource intensive, and are likely to have the greatest uncertainty.

1  The following eight chemicals or chemical classes and their associated effects were chosen for
2  prototype development:

3  • Tier 3:
4        o   Benzene and leukemia (molecular epidemiology),
5        o   Ozone and lung inflammation and injury (molecular clinical studies), and
6        o   Benzo[a]pyrene (B[a]P)/polycyclic aromatic hydrocarbons (PAHs) and liver cancer
7            (molecular clinical studies meta-analyses and *in vivo* rodent bioassay).

8  • Tier 2:
9        o   Chemicals associated with diabetes and obesity ("big data" knowledge mining),
10       o   Chemicals associated with thyroid hormone disruption (short duration *in vivo*
11           exposure bioassays-alternative species), and
12       o   Chemicals associated with cancer (short duration *in vivo* exposure bioassays-
13           mammalian).

14 • Tier 1:
15       o   Chemicals associated with cancer and noncancer disorders especially
16           developmental (QSAR) and
17       o   Chemicals associated with thyroid hormone disruption (high throughput *in vitro*
18           assays).

19 Highlighted methods include molecular clinical and epidemiologic studies, *in vivo* molecular
20 nonhuman studies, high-content *in vivo* assays (mammalian and nonmammalian species),
21 bioinformatics, data mining, high-throughput *in vitro* screening assays, and quantitative structure
22 activity modeling. NexGen methods and results were compared to robust traditional data set
23 results.

24 Both bottom-up and top-down perspectives were used to evaluate the available data. The top-down
25 approach focuses on higher system-level indicators of disease resulting from environmental
26 exposures to known chemicals based on data from human clinical and epidemiologic studies. The
27 bottom-up approach focuses on information describing chemically induced alterations in molecular
28 and cellular components, as well as their network interactions. These data support the capability to
29 develop risk assessments for chemicals with little or no traditional data. Additionally, these data
30 can further inform assessments based on traditional data.

31 Data and insights from both bottom-up and top-down approaches are integrated to inform
32 understanding of potential health risks associated with chemical exposures. The following
33 summarizes lessons learned from development of the prototypes, as well as challenges for
34 incorporating novel data streams to inform risk assessment:

35 • Advances in genomics, epigenomics, transcriptomics, metabolomics, and cell and systems
36   biology, together with advanced analytical methods in biostatistics, bioinformatics, and
37   computational biology, have the potential to increase dramatically understanding of the
38   molecular basis of disease and environmental factors that alter disease risks.

39 • Of particular importance are the many new tools that facilitate testing and evaluation, on an
40   unprecedented scale, of chemicals with limited or no traditional data. ToxCast™ and
41   Toxicology in the 21st Century (Tox21) Programs provide examples.

1 • New approaches can be used to identify biological patterns or signatures that are associated
2 with specific diseases, thus facilitating grouping and evaluating chemicals based on the
3 mechanistic underpinnings of specific diseases. The Comparative Toxicogenomic Database
4 provides a partial example.

5 • These signatures are best developed and understood as they relate to apical outcomes using
6 systems biology. Conceptualization of these relationships among early molecular events,
7 intermediate events, and apical outcomes are often termed mode of action or "adverse
8 outcome pathways."[1]

9 • Signatures appear exposure-dose dependent (i.e., the magnitude of response changes with
10 changes in exposure-dose) and hence, might be used to prioritize chemicals based on relative
11 potencies, to serve as biomarkers of exposure and effect, and to inform quantitative risk
12 assessment. Biological processes also are often time-dependent, which can complicate
13 interpretation.

14 • The links between molecular perturbations and disease outcomes are influenced by a number
15 of variables, that is, metabolism, cell type, genomic variants, cell and tissue interactions, and
16 species. Thus, some test systems might better predict the potency of a chemical to disrupt
17 normal biology than predict the specific adverse outcome resulting from that disruption.

18 • Historically, many controversial risk assessment issues lack data for substantive progress in
19 understanding. NexGen approaches can provide new data types to improve the
20 characterization of human variability and susceptibility, cross-species relevance, and low
21 exposure-dose-response relationships via understanding mechanistic commonalities and
22 differences. These issues are discussed in this report.

23 The prototype results presented in this report demonstrate proof-of-concept for an integrated
24 approach to risk assessment based on molecular, computational, and systems biology. In addition,
25 they explore which types of information appear most valuable for specific purposes and articulate
26 some decision considerations for use of data. Based on lessons learned from this effort, near-term
27 and longer term implications for risk assessment are also discussed.

28 Further advances in methods and knowledge undoubtedly will occur over the near term. Logistical
29 and methodological challenges in interpreting and using newer data and methods in risk
30 assessment, however, remain significant. Hence, incorporating new information into risk
31 assessment will remain an ongoing opportunity.

---

[1]An adverse outcome pathway has been defined as the mechanistic or predictive relationship between initial chemical-biological interactions (i.e., molecular initiating event[s]; [MIE]) and subsequent perturbations to cellular functions sufficient to elicit disruptions at higher levels of organization, culminating in an adverse phenotypic outcome in an individual and population relevant to risk assessment (e.g., disease progression or organ dysfunction in humans) (Ankley, G. T. et al. 2010). Although commonly used, the term is something of a misnomer; pathways are not intrinsically adverse or nonadverse but rather pathways when perturbed in specific ways can lead to adverse outcomes. The same can be said of the commonly use term "toxicity" pathways.

[page intentionally left blank]

*This document is a draft for review purposes only and does not constitute Agency policy. DRAFT – Do Not Cite or Quote.*

September 2013                                                   xiii

# 1. Introduction

**Box 1. Next Generation Risk Assessment (NexGen)**

This report describes the NexGen program, a multiyear, multi-organization effort to develop and evaluate new molecular, computational, and systems biology informed approaches to risk assessment. The goal of this effort is to advance risk assessment by facilitating faster, less expensive, and more robust assessments of public health risks by EPA's Office of Research and Development. The specific aims of the program are to:

- demonstrate proof of concept that recent advances in biology can better inform risk assessment;
- understand what information is most useful for particular purposes (value of information);
- articulate decision considerations for use of new types of data and methods to inform risk assessment; and
- identify important data gaps.

1  In recent years, public concern has grown
2  about the number of chemicals in the
3  environment and the ability to assess the
4  risk to human health from potential
5  exposures. Efforts by government agencies,
6  including the U.S. Environmental
7  Protection Agency (EPA), to protect public
8  health from unreasonable chemical
9  exposure have been hindered by
10  limitations in current chemical testing
11  methods and data. The European
12  Commission has underscored this concern
13  with recent initiatives to identify the many
14  thousands of largely untested chemicals in
15  use today and to increase the available
16  toxicity information on those chemicals relative to the amount of chemical used and the potential
17  for exposure in the environment (ECHA 2013a). As a result, significant efforts are underway
18  throughout the world to redesign toxicity testing and understand how advances in biology,
19  biotechnology, and computational science during the past two decades can be used in risk
20  assessment. Specific goals are to increase dramatically our ability to test and assess chemicals more
21  rapidly, understand disease processes and relationships to environmental factors, and facilitate the
22  process from data acquisition to data analysis.

23  The technologies that have emerged from the sequencing of the human genome have ushered in a
24  new era in biology (Collins, FS 2010) that supports the above goals. Advances in genomics,
25  epigenomics, transcriptomics, metabolomics, proteomics, and cell and systems biology,[2] together
26  with advanced analytical methods in biostatistics, bioinformatics, and computational biology, have
27  dramatically increased our understanding of the molecular basis of disease—what causes disease
28  and what exacerbates and ameliorates our risk of disease. Molecular signatures and other
29  biomarkers are helping identify and define disease states and responses, and thousands of
30  variations in previously unknown human health risk factors are being identified.

31  Researchers are generating massive amounts of biological data from the new "omics" technologies.
32  Approximately 1.8 zettabytes ($10^{21}$) of new data are generated every year, roughly doubling the
33  world's information resource every 2 years (Dearry 2013). More than 50,000 "genomics" papers
34  are published each year (NCBI 2013). Large, publicly available data sets now support analyses of
35  environmental health data on an unprecedented scale, driving further discovery of new knowledge
36  (Dearry 2013, Abecasis et al. 2012, ENCODE Project Consortium 2012, Mechanic et al. 2012, Wang, I
37  et al. 2012, Collins, MA 2009, Thomas, RS et al. 2009, Ramasamy et al. 2008). Concomitantly,
38  powerful data mining, statistical, and bioinformatics methods have been developed to identify,

---

[2]**Systems biology** is defined as a "scientific approach that combines the principles of engineering, mathematics, physics, and computer science with extensive experimental data to develop a quantitative as well as a deep conceptual understanding of biological phenomena, permitting prediction and accurate simulation of complex (emergent) biological behaviors" (Wanjek 2013). See Wanjek's (2013) Web article *Systems as Biology as Defined by NIH* for more discussion of systems biology.

*This document is a draft for review purposes only and does not constitute Agency policy. Do not cite or quote.*

September 2013                                                    1

1  prioritize, and classify biomarkers with high discriminatory ability (Fang et al. 2012), and to store
2  and manage the information in database libraries, including the Gene Expression Omnibus (GEO)
3  (NCBI 2012a), the Kyoto Encyclopedia of Genes and Genomes (KEGG) (Kanehisa Laboratories
4  2013), the Comparative Toxicogenomic Database (CTD) (NIEHS 2013), and the Epigenomics
5  Database (Chadwick 2012, NCBI 2009). As integration across differing types of data and levels of
6  biological organization occurs (Birney 2012, ENCODE Project Consortium 2012), the degrees to
7  which environmental risk assessment will be transformed and our understanding of disease at the
8  individual and population level will be advanced are anticipated to be significant (Bhattacharya et
9  al. 2011, Chiu et al. 2010).

10  Scientific discovery is now moving away from the traditional approach of individual scientists'
11  conducting experiments in their laboratories to pooling of data into publicly available databases
12  and broad collaborative participation in problem solving (Dearry 2013, Friend 2013, Derry et al.
13  2012). The magnitude of changes was highlighted in
14  remarks by Frances Collins (Director of National
15  Institutes of Health [NIH]) who stated that within the
16  near future, most people in the United States will have a
17  genome scan in their medical records as a tool for
18  diagnosis, prognosis, and treatment of disease (Collins,
19  FS 2010).

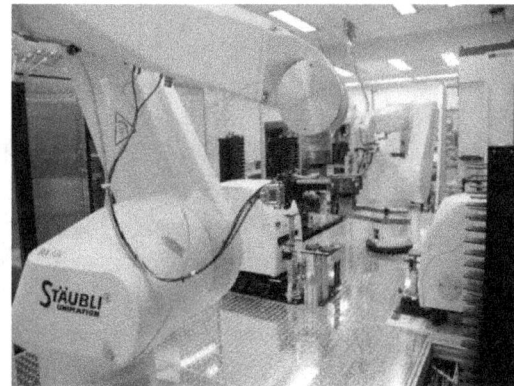

20  The impact of recent scientific advances on our ability
21  to conduct risk assessments and protect public health
22  cannot be overestimated. Particularly relevant to
23  environmental risk assessment is that new data types
24  and methods will result in much more rapid evaluation
25  of chemicals, increase identification of causal
26  mechanisms of disease, and provide a more profound
27  understanding of the interrelated roles of genetics,
28  epigenetics, and environmental factors. Experiments
29  already can be conducted much more rapidly and
30  efficiently using robotics and *in vitro* assays to measure
31  molecular functions. Two examples are (1) Toxicology
32  in the 21st Century (Tox21) testing of 10,000 chemicals
33  within 3 years using approximately 150 assays (Figure

Figure 1. Toxicology in the 21st Century (Tox21) robot conducts bioassays on 10,000 chemicals. A robot arm (foreground) retrieves assay plates from incubators and places them at compound transfer stations or hands them off to another robot arm (background) that services liquid dispensers or plate readers. Photo by Maggie Bartlett (NHGRI 2012).

34  1) (Tice et al. 2013), and (2) the study of gene- and environment-wide associations with disease in
35  tens of thousands of humans with multiple diseases—both unimaginable feats 15 years ago (Friend
36  2013, Mechanic et al. 2012). With the burgeoning amounts of data produced by high-volume testing
37  and discovery, an effort in the European Union called "Safety Evaluation Ultimately Replacing
38  Animal Testing," SEURAT-1 (http://www.seurat-1.eu/) has begun to develop a conceptual
39  framework that can be used as a basis to combine information derived from predictive tools to
40  support a safety assessment process. The overarching SEURAT-1 research strategy is to adopt a

1    toxicological mode-of-action[3] approach to describe how any substance might adversely affect
2    human health, and to use this knowledge to develop complementary theoretical, computational,
3    and experimental (*in vitro*) models.

4    In collaboration with its partners (see Acknowledgments), EPA initiated the NexGen program in
5    2010 to evaluate the use of these recent advances in biological and computational sciences for risk
6    assessment (Text Box 1). We initially conducted workshops and solicited expert opinion to develop
7    a framework and suggestions for prototype assessments that address the needs of the public and
8    the risk assessment community. Federal, state, and other partners participated in the workshops
9    and continue to provide advice, data, and review for NexGen reports. Text Box 2 lists related,
10   ongoing legislation and government research activities in Europe and the United States.

---

[3]"Mode-of-action" is one term used to reference a mechanistic understanding of the impact of a chemical on human health. Other terms include "disease signature" and "network perturbations" from epidemiology for example, while toxicologists might reference the same concept using the terms "toxicity pathway," "mode-of-action," or "adverse outcome pathway." In general, this report uses the term "mechanism of action," in accordance with the National Research Council (NRC) report, *Science and Decisions: Advancing Risk Assessment* (2009); however, the exact term used in a specific section of this report is based on the references used and the context of the discussion.

**Box 2. Current Legislation and Governmental Research Activities in Europe and the United States**

**EUROPEAN LEGISLATION AND ACTIVITIES**

In response to environmental concerns, a desire for increased assessment efficiencies, and a desire to reduce reliance on *in vivo* animal testing, the European Union (EU) enacted an expansive new program called Registration, Evaluation, Authorisation and Restriction of Chemicals (REACH) in June 2007. This legislation places greater responsibility on industry to test and manage the risks posed by their chemicals. Under REACH, companies must develop detailed technical dossiers and chemical safety reports and submit them to the European Chemicals Agency (ECHA). Approximately 12,000 chemicals have been registered for consideration with ECHA. Many more chemicals are anticipated in the near future. Additionally, the 7th Amendment to the EU Cosmetics Directive prohibits putting animal-tested cosmetics on the market in Europe after 2013. Although current alternative methods more closely resemble traditional methods, the EU has invested 50M Euros in a research program to further next-generation methods (OECD 2012). Current ECHA guidance is available on the use of quantitative structure-activity relationships (QSARs), *in vitro* assays, and read-across (also known as near-analog structure-activity relationships) to support assessments.

REACH and the 7th amendment will significantly impact nearly all multinational companies and are important drivers for the development and use of new molecular-based methodologies. Europe's chemical trade accounts for about 40% of the global market, involving 27 countries and almost half a billion people.

**The Joint Research Centre (JRC)** is the scientific and technical arm of the European Commission. It provides scientific advice and technical support to EU policies. The JRC has seven scientific institutes (featuring laboratories and research facilities) located at five sites: Belgium, Germany, Italy, the Netherlands, and Spain. The JRC's Institute for Health and Consumer Protection's main research relevant to NexGen includes integrated risk and benefit assessments of chemical substances; fit-for-purpose analytical tools to help ensure the safety of food and consumer products; and optimization and validation of methods that reduce the reliance on animal tests in the safety assessment of chemicals.

**U.S. ACTIVITIES**

Several documents have guided the NexGen effort, including the Strategic Plan for the Future of Toxicity Testing and Risk Assessment at the U.S. Environmental Protection Agency (EPA 2009a), the Toxicology in the 21st Century (Tox21) strategy, and the National Institutes of Health Strategic Plan (NIEHS 2012c). Ongoing research activities of several federal agencies that have informed and continue to inform the NexGen effort are described below.

**The Centers for Disease Control and Prevention (CDC)** has several groups involved in systems biology and computational environmental health and occupational research. The **National Center for Environmental Health (NCEH)** and **Agency for Toxic Substances and Disease Registry (ATSDR)** scientists in the Computational Toxicology Laboratory have applied several new approaches for improving chemical risk assessments. They have mined the National Health and Nutrition Examination Survey (NHANES) data set to obtain high-quality analytical and human health information, which is representative of the general U.S. population, and used computer modeling to identify sensitive populations for health outcomes at environmental exposure levels. A second project involved use of NHANES public health genomics data to identify allelic differences in ALA dehydratase for susceptibility to lead-induced hypertension. Another concerned the development and application of QSAR, physiologically based pharmacokinetic (PBPK), and molecular docking approaches. These studies involved both data mining of the published scientific literature and collaborative laboratory studies with scientists at the Food and Drug Administration (FDA).

**The National Institute for Occupational Safety and Health (NIOSH)** is investigating susceptibility gene variants that contribute to the development and severity of occupational diseases using high-density and high-throughput (HT) genotyping platforms. Understanding the genetic contribution to the development, progression, and outcomes of complex occupational diseases will help improve the accuracy of risk assessment and improve safe exposure levels for genetically susceptible groups in the workforce.

**The FDA National Center for Toxicological Research (NCTR)** is conducting translational research to develop a scientifically sound basis for regulatory decisions and reduce risks associated with FDA-regulated products. NCTR research evaluates biological effects of potentially toxic chemicals, defines the complex mechanisms that govern their toxicity, identifies the critical biological events in the expression of toxicity, discovers biomarkers, and develops new scientific tools and methods to improve assessment of human exposure, susceptibility, and risk. Examples of tools created by NCTR include ArrayTrack™, Decision Forest, Endocrine Disruptor Knowledge Base (EDKB), Gene Ontology for Functional Analysis (GOFFA), and SNPTrack. Efforts include the MicroArray Quality Control (MAQC) consortia.

**Box 2. Current Legislation and Governmental Research Activities in Europe and the United States (Continued)**

U.S. ACTIVITIES (CONTINUED)

**The National Institutes of Health (NIH) National Center for Advancing Translational Sciences (NCATS)** conducts research to resolve scientific and technical challenges that might cause barriers to the efficient development of new treatments and tests to improve human health. The National Chemical Genomics Center (NCGC) at the National Center for Advancing Translational Sciences applies high-throughput screening (HTS) assay guidance, informatics, and chemistry resources for NCAT's Re-engineering Translational Sciences research projects. Specifically, NCGC research programs include assay development and HTS, and participation in Tox21. NCGC Assay Biology Teams are researching optimization of biochemical, cellular, and model organism-based assays submitted by the biomedical research community for HT small molecule screening. The results of these screens (probes) can be used to further examine protein and cell functions and biological processes relevant to physiology and disease (NIH 2012).

**The National Human Genome Research Institute (NHGRI)** was established by NIH in 1989 to implement the International Human Genome Project to map the human genome. NHGRI has developed programs for a variety of research projects including Encyclopedia of DNA Elements (ENCODE), Gene Expression Omnibus (GEO), and collaborative projects, including the Comparative Toxicogenomic Database (CTD), HapMap, and Gene. Through the application of these tools, NHGRI hopes to gain a greater understanding of human genetic disease, and develop better methods for the detection, prevention, and treatment of genetic disorders.

**The National Institute of Environmental Health Science (NIEHS) and the National Toxicology Program (NTP)** have played an integral role in the development and application of HTS data. Current research is focused on developing and validating Tox21 approaches to improve hazard identification, characterization, and risk assessment (Birnbaum 2012, Serafimova et al. 2007). The NTP HTS program has three specific goals: (1) prioritizing substances for in-depth toxicological evaluation, (2) identifying mechanisms of action for further investigation (e.g., disease-associated pathways), and (3) developing predictive models for *in vivo* biological response (i.e., predictive toxicology). NTP is developing innovative and flexible approaches to data integration, both across research programs and across different data types (e.g., HT, mechanistic, animal studies) (Bucher et al. 2011). These efforts seek to integrate results from new techniques with traditional toxicology data to provide a public health context.

**The Engineer Research and Development Center (ERDC), the research organization of the U.S. Army Corps of Engineers**, conducts research and development in support of warfighters, military installations, and civil works projects involving water resources and environmental missions. The ERDC Toxicogenomics research cluster focuses on using genomics to develop tools to rapidly assess toxicity of military chemicals in a wide range of animals, identifying gene biomarkers of exposure, understanding the mechanisms by which military chemicals cause toxicity, and extrapolating toxicity effects across multiple species. Capabilities of the team include advanced instrumentation to characterize impacts of chemicals on gene expression with high-density gene arrays, DNA sequencing, and real-time polymerase chain reaction (RT-PCR) assays. ERDC Toxicogenomic projects include development of rapid assays to assess whole genome impacts of munitions-related compounds, including gene arrays with short exposure screening in daphnia, rat cells, rat livers, and fish; comparison of genomic and behavioral responses of fathead minnows and zebrafish to chemical exposures; conservation of response to nitroaromatics across species; and support for a toxicogenomic assessment framework to integrate predictive toxicology of munitions-related compounds.

**Several EPA Office of Research and Development laboratories and centers** have been involved in NexGen. EPA's **National Center for Environmental Assessment (NCEA)** has assumed a leadership and coordination role for the NexGen effort. The **National Center for Computational Toxicology (NCCT)** is the largest component of EPA's Computational Toxicology Research Program. The Center coordinates computational toxicology research on chemical screening and prioritization, informatics, and systems modeling. NCCT research includes the (1) use of informatics, HTS technologies, and systems biology to develop accurate and flexible computational tools that can screen the thousands of chemicals for potential toxicity; and (2) application of mathematical and advanced computer models to help assess chemical hazards and risks. EPA's **National Center for Environmental Research (NCER)** supports extramural computational toxicology research. The **National Health and Environmental Effects Research Laboratory (NHEERL)** conducts toxicological, clinical, and epidemiological research to improve the process of human health risk assessments, including development of biological assays and toxicological assessment methods, predictive pharmacokinetic/pharmacodynamic models, and advanced extrapolation methods.

*This document is a draft for review purposes only and does not constitute Agency policy. Do not cite or quote.*

September 2013                                                    5

1　The initial NexGen prototypes were designed to provide concrete examples that illustrate the
2　potential for various new methods and data to be used for specific risk assessments within a
3　decision context[4] and to foster further discussion in the risk assessment and risk management
4　communities to promote continual improvement.

5　This report presents and discusses the results of this effort, and is organized as follows:

6　• Section 2: Preparation for Prototype Development – describes the preliminary work and
7　workshops conducted to characterize the decision context and conceptual framework and to
8　identify the stakeholders and key issues so that the prototypes provide examples relevant to
9　the needs of the risk assessment community.

10　• Section 3: The Prototypes – provides detailed examples of the use of various advanced
11　methods and data in each of the three tiers, starting with chemicals for Tier 3 "Major Scope
12　Assessments," which are data-rich chemicals, proceeding to Tier 2 and Tier 1 chemicals that
13　have increasingly limited or no *in vivo* data sufficient to conduct a traditional (e.g., IRIS) risk
14　assessment.

15　• Section 4: Advanced Approaches to Recurring Issues in Risk Assessment – describes how
16　advanced methods are being used to address recurring and challenging issues, including
17　characterizing variability in deriving toxicity values and assessing potential hazards from
18　exposure to mixtures.

19　• Section 5: Lessons Learnedfrom Developing the Protypes – describes lessons learned in
20　developing the Tier 1, 2, and 3 prototype examples.

21　• Section 6: Conclusions– outlines the major challenges and future direction for the NexGen
22　program.

23　• Appendix A lists technical papers supporting this report.

24　• Appendix B provides a glossary.

## 2. Preparation for Prototype Development

### 2.1. Consideration of Decision Context

25　One of the first tasks undertaken in planning the NexGen effort was consideration of the various
26　environmental situations of concern to EPA's Program Offices—in other words, the decision context
27　[termed in Cote et al. (2012) and the National Research Council (NRC 2009) and National Academy
28　of Science (NAS 2007) reports]. Decision context defines what environmental management decision
29　is being made and why, as well as its relationship to other decisions previously made or anticipated.
30　EPA Program Offices are generally organized around specific pieces of environmental legislation,
31　such as the Clean Air Act and the Clean Water Act, and are responsible for administering those laws.
32　Each major piece of legislation brings different responsibilities and nuances to problems faced by
33　risk managers. In Figure 2, the decision context is represented in three categories for ease of
34　description. The characteristics that define the three decision context categories and examples of
35　specific problems faced by the Program Offices are shown. This figure elaborates on the decision

---

[4]See Section 2.1 for a definition of "decision context" and a discussion of its use.

1    context figure in the report, *Science and Decisions: Advancing Risk Assessment,* and is the result of
2    discussions with EPA Program Offices (EPA 2011b).

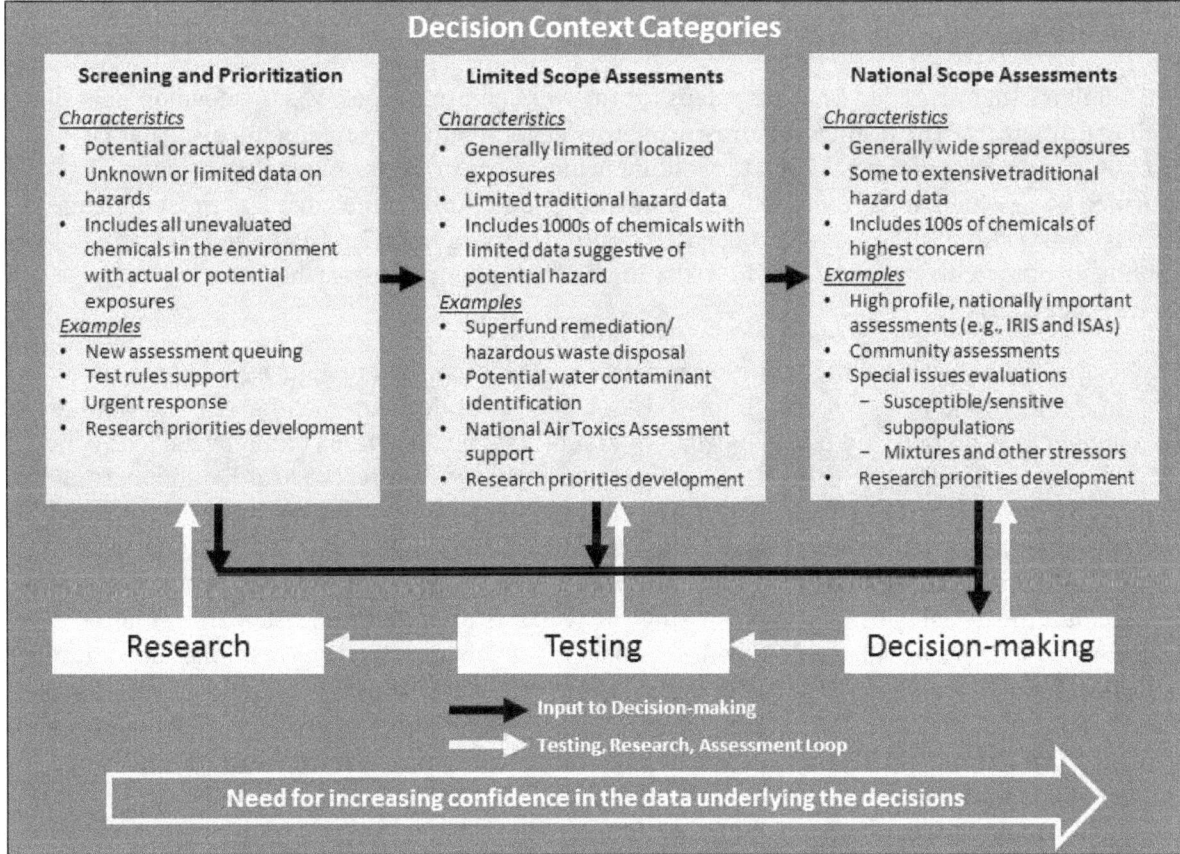

Figure 2. Description of decision context categories provided by EPA Program Offices. These decision context categories reflect the range of environmental problems to be addressed, from the need to screen many untested chemicals in the environment to national regulations for high profile chemicals. The flow from decision context through risk assessment to decision-making and the related roles of testing and research are also noted.

3    Three factors integral to the decision context for risk managers are the potential exposure, the
4    number of chemicals that should be considered, and the weight of scientific evidence for supporting
5    decision-making. Both legislative language and the history of specific regulatory programs
6    influence the numbers of chemicals considered and the uncertainty in supporting data that can be
7    tolerated. Tier 3 decision context focuses on nationally relevant chemicals with widespread
8    exposures and established hazards and for which major regulatory evaluations are likely in
9    progress. An example would include the International Agency for Research on Cancer (IARC)
10   benzene assessment (2012) where molecular mechanistic information was used to support the
11   causal link between benzene and hematopoietic cancers, particularly when the epidemiology data
12   were somewhat limited. Tier 2 focuses on chemicals for which exposure or hazard appears limited
13   or available data for detailed assessment are limited. An example includes evaluation of biological
14   activity and cumulative risk potential of conazole fungicides (EPA 2011e) and potential endocrine
15   disruptors (EPA 2011c), both of which are based on molecular biology data in combination with
16   traditional data. Tier 1 decision context focuses on the tens of thousands of chemicals present in

1  commerce in significant amounts, but for which we have little knowledge of exposure levels or
2  potential health effects. An example is the high-throughput (HT)-based evaluations of Deep Water
3  Horizon Gulf oil spill dispersants (Judson et al. 2010).

## 2.2. A Framework

4  A second task that preceded finalizing plans for the NexGen prototypes was the development of a
5  guiding framework. The framework draws together several important elements of earlier risk
6  assessment frameworks and articulates guiding principles for risk assessment development
7  informed by new data types and methods. A draft version of this framework was presented and
8  discussed in October 2010 at a meeting with scientific experts (EPA 2010) and in February 2011 at
9  a public meeting with stakeholders (EPA 2011a). The framework is described in a report by
10 Krewski et al. (2013).

11 The NexGen framework is built on three cornerstones (as illustrated in Figure 3): (1) new risk
12 assessment methodologies to better inform risk management decision-making; (2) new data types
13 from advances in biology and toxicology on understanding perturbations of biological pathways;
14 and (3) a population health perspective that recognizes that most adverse health outcomes involve
15 multiple determinants. The NexGen framework integrates these three cornerstones into a
16 framework for risk science that progresses in three phases: (1) Objectives, (2) Risk Assessment,
17 and (3) Risk Management. Phase 1 (Objectives) focuses on problem formulation and scoping, taking
18 into account the decision context and the range of available or admissible risk management
19 decision-making options. Phase 2 (Risk Assessment) seeks to identify disease or outcome pathways
20 using new toxicity testing tools and technologies and attempts to improve the characterization of
21 risks and uncertainties using advanced risk assessment methodologies. Phase 3 (Risk Management)
22 involves the development of evidence-based population health risk management strategies of a
23 regulatory, economic, advisory, community, or technological nature, based on sound principles of
24 risk management decision-making. Implementation of the NexGen framework is exemplified with a
25 series of case-study prototypes, illustrating how aspects of the framework have been put into
26 practice.

27 NRC provided a blueprint for pathway-based toxicity testing in its 2007 report, *Toxicity Testing in*
28 *the 21st Century: A Vision and a Strategy* (NRC 2007). Guidance on some of the new risk assessment
29 methods is provided by the 2009 report, *Science and Decisions, Advancing Risk Assessment*
30 [NRC (2009)]. The integration of a population health approach was drawn from the McLaughlin
31 Centre's integrated risk management and population health framework. Key elements of risk
32 science and population health are combined to offer a multidisciplinary approach to the assessment
33 and management of health risk issues (Krewski et al. 2007).

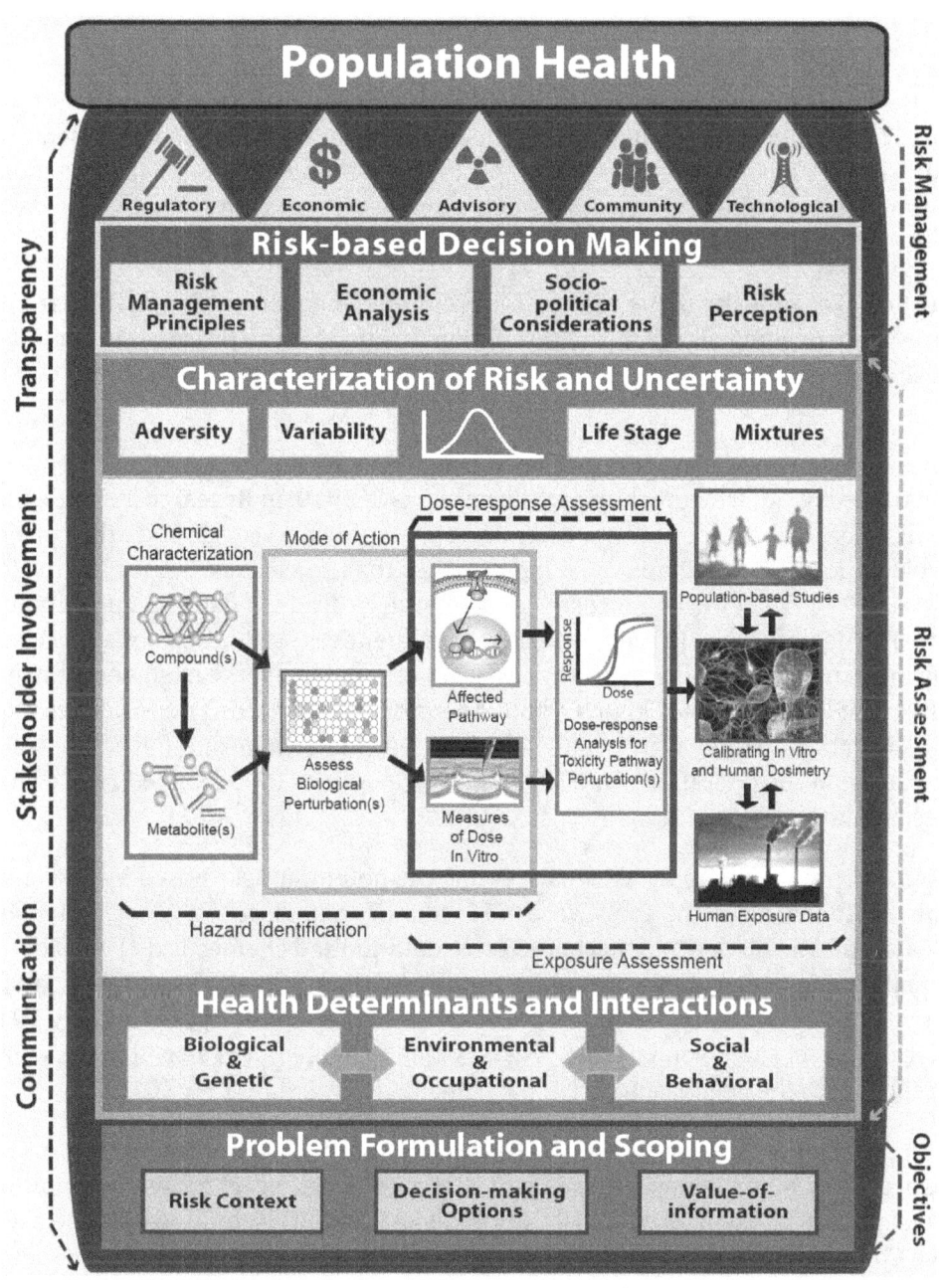

Figure 3. The Next Generation Framework for Risk Science. This framework is divided into three phases: (1) Objectives: Problem Formulation and Scoping takes into consideration Risk Context,[5] Decision-Making Options, and Value of Information; (2) Risk Assessment involves three sub categories: (A) Health Determinants and Interactions, (B) New Scientific Tools and Technologies, and (C) New Risk Assessment Methodologies; and (3) Risk Management involves two categories: (A) Risk-Based Decision-Making that involves Risk Management Principles, Economic Analysis, Socio-Political Consideration, and Risk Perception, and (B) Risk Management Interventions with five possible categories: Regulatory, Economic, Advisory, Community, and Technical (Krewski et al. 2013).

---

[5]The term decision context is used elsewhere in this report.

---

*This document is a draft for review purposes only and does not constitute Agency policy. Do not cite or quote.*

September 2013                                    9

## 2.3. Science Community and Stakeholder Engagement

1  Outreach to the science community and stakeholder groups was part of the NexGen strategy from
2  its inception. This document in its final form is viewed as an interim step to implementation of new
3  advances in risk assessment and is intended to promote further discussion with stakeholders
4  toward continual improvement of risk assessments and prototypes informed by new data types and
5  methods.

6  Given the technical complexity of the research, stakeholder engagement is a particular challenge
7  and will necessitate ongoing outreach and discussion throughout the process. Our initial efforts are
8  described below.

### 2.3.1. Expert Workshop

9   EPA convened a 3-day expert workshop on November 1–3, 2010, in Research Triangle Park, North
10  Carolina, to discuss the draft framework, early draft prototypes, research, and other project
11  elements. The workshop sought individual input, rather than consensus, in meeting its discussion
12  goals. Days 1 and 2 of the workshop focused on deliberative drafts of data-rich prototype health
13  assessments. The goals were to (1) refine health assessment case studies of data-rich chemicals
14  informed by molecular biology (i.e., "prototypes"); (2) enhance "reverse engineering" from
15  molecular system biology data, to "known" public health risk estimates based on *in vivo* human and
16  animal bioassay data to demonstrate proof of concept, elucidate value of information, and
17  characterize decision considerations; and (3) summarize options for expanded future work and
18  research needs.

19  Day 3 focused on approaches applicable to assessing the potential risks posed by chemicals with
20  limited or no traditional data. The goals were to (1) identify and discuss a wider variety of new data
21  types, methods, and knowledge to help characterize data-limited chemicals; (2) consider how this
22  information might augment, extend, or replace traditional data in health assessment; and
23  (3) summarize options for expanded future work and research needs. Approximately 40 federal
24  and nonfederal experts and 80 and partner organization staff members attended the workshop. A
25  workshop report with the agenda and list of participants is available (EPA 2010).

26  In 2012, both the Science Advisory Board (SAB) and the Board of Scientific Counselors (BOSC)
27  reviewed aspects of the NexGen program as part of their evaluations of EPA's computational
28  toxicology research (SAB 2013, BOSC 2010). Both the SAB and BOSC commended the exceptional
29  efforts of the Computational Toxicology Research Program to advance hazard/risk assessment and
30  provided recommendations for the continued success of the program. The reviews emphasized
31  further research on chemical exposure pathways resulting from human activity patterns (e.g.,
32  ExpoCast); engagement of the scientific community and stakeholders to foster future partnerships
33  and promote information exchange; broader outreach for dissemination of scientific findings;
34  gathering of user-feedback from the general public; improvements in data access through enhanced
35  website navigation; and development of guidelines for data usage.

### 2.3.2. Stakeholder Involvement

**Stakeholder Public Dialogue Conference**

1 To engage stakeholders in the early stages of the NexGen program, EPA sponsored a public dialogue
2 conference, "Advancing the Next Generation of Risk Assessment," on February 15 and 16, 2011, in
3 Washington, DC. This conference presented stakeholders with an opportunity to learn about the
4 NexGen program, and to provide their thoughts on the challenges the project faces and its path
5 forward. Approximately 160 participants, representing 11 stakeholder groups (Figure 4), attended
6 the conference. The conference report includes the agenda, list of participants, and
7 recommendations of the group (EPA 2011a). In addition to this conference, "one-on-one"
8 interviews (described below) were conducted with leaders of public-interest groups and the
9 business community.

**Public Interest Group Perspectives**

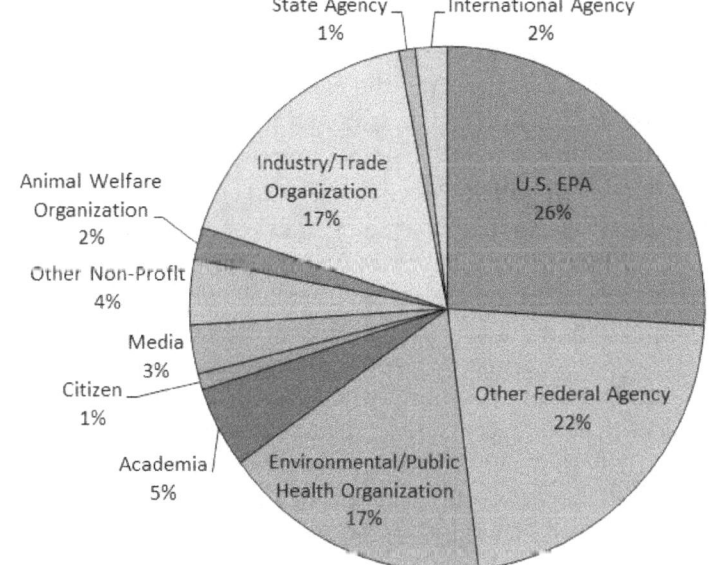

10 After the work shop, follow-up
11 informal one-on-one interviews
12 were conducted in mid-2010
13 with several Washington,
14 DC-based representatives of
15 national environmental, public
16 health, and animal welfare
17 public-interest organizations.
18 Ronald White, a faculty member
19 at Johns Hopkins Bloomberg
20 School of Public Health,
21 conducted these interviews and
22 informational meetings as a
23 component of his research on
24 public engagement regarding
25 emerging risk assessment
26 methods. He also developed a

Figure 4. Categories of stakeholders that attended the February 2011 NexGen public dialogue conference (EPA 2011a).

27 web-based assessment in late 2010 to ascertain, from nongovernmental public-interest
28 organizations, their knowledge and interest in emerging scientific approaches for
29 chemical/pollutant risk assessment. Of the 24 organizations contacted, 8 (33%) responded to the
30 assessment.

31 A key question raised in these interviews and web-based assessment was how relevant the NexGen
32 program is to near-term EPA pollutant/chemical risk assessment procedures and control policies.
33 The public-interest stakeholders interviewed and those who responded to the online assessment
34 questions generally supported the concept of integrating the results from emerging biological
35 science and analytic techniques into EPA's approach to conducting chemical health-based risk
36 assessment. Significant concerns emerged, however, regarding the following:

1   • The potential for overstating the utility of NexGen approaches.

2   • How NexGen prototypes will address key risk assessment methodological issues, such as low-
3     dose exposure assessment, population variability in response, and additivity to background
4     exposures and disease processes.

5   • The transparency of the NexGen assessment development process and opportunities for
6     early, meaningful engagement by public-interest organizations, and the application of NexGen
7     approaches in risk management.

### Business Community Perspectives

8   Industry or business perspectives on NexGen approaches also were of interest. Dr. Gerald Poje,
9   (Environmental Health Consultant, Former Board Member of the U.S. Chemical Safety and Hazard
10  Investigation Board) had follow-up discussions with nine individuals representing the high-
11  production volume and smaller specialty chemical manufacturing industries, the pharmaceutical
12  industry, the retail sector, and the energy sector. The participants were generally optimistic about
13  potential advances in risk assessment and identified two potential advantages: (1) better prioritize
14  the needs for more expensive and longer duration whole-animal testing, and (2) save time and
15  money while rationalizing decisions in a tier-based manner using HT and other Tier 1 and Tier 2
16  tests. They also suggested that the success of NexGen effort depends on EPA's ability to prove the
17  value of the newer, tiered approach within EPA's emerging risk assessment model, the level of
18  EPA's investment in the long-term iterative NexGen research effort, and the timely and effective
19  communication of the evidence to support science-based risk assessment.

20  Some in the business community expressed concern over whether EPA could match the
21  pharmaceutical industry's growing infrastructure (needed to support and sustain a NexGen-like
22  effort) such as EPA's ability to unite sufficient numbers of expert biologists, chemists, and
23  bioinformatics to guide the program to a successful conclusion. The technical complexity of the
24  NexGen program might also hinder its impact on current risk assessment, risk management, and
25  business development practices, given the many unknowns that remain. Cultural challenges in
26  winning over a larger community, who will welcome the use of more recent advances in risk
27  assessment methods; however, was thought to be surmountable if EPA could be effective at
28  capacity building and communicating how new data types and approaches could be used for risk
29  assessment.

## 2.4. Recurring Issues in Risk Assessment

30  The fourth task that preceded the actual prototype development was identification of problematic
31  issues that might be substantively informed by new methods and data. The issues included problem
32  formation, adversity classifications and weight of evidence, dose-response modeling (especially at
33  the low-dose end), variability in human response (due to a variety of factors), interspecies
34  extrapolation, mixtures risk assessment, and characterization of uncertainty. These issues are
35  explored in the prototypes to the extent feasible, and some are discussed in more detail in papers
36  on human variability (Zeise et al. 2012), early-life exposure and later-life disease risks (Boekelheide
37  et al. 2012), and multifactorial interactions of environment and genes (Patel et al. 2012a, Patel et al.
38  2012b, Zhuo et al. 2012, Shen et al. 2011, Smith, MT et al. 2011).

# 3. The Prototypes

1   EPA's Office of Research and Development, in conjunction with other federal, state, academic,
2   public, and private partners (see Acknowledgments), developed prototype assessments to provide
3   concrete illustrations of how new and emerging information could inform risk assessment. The
4   prototypes used a variety of study types, methods, data, and risk assessment approaches, and are
5   intended to (1) engender movement in the field of risk assessment from strategy to practical
6   application of new approaches, and (2) foster discussion and refinement of approaches in the risk
7   assessment and risk management communities, as
8   well as with the public.

9   The results presented in this report demonstrate
10  proof of concept, provide insight on what types of
11  information are valuable for specific purposes, and
12  provide examples of the decision considerations for
13  reasonable, consistent, and coherent use of the new
14  types of information for specific applications. The
15  prototypes also illustrate many of the challenges.
16  Text Box 3 lists selection criteria used in choosing
17  prototypes. Figure 5 broadly categorizes the types of
18  methods aligned to decision context and evaluated
19  in the prototypes. As noted earlier, the number of
20  chemicals that need to be evaluated and the level of
21  confidence required for decision-making are key
22  components of designing fit-for-purpose
23  assessments. The integration of knowledge from a
24  wide variety of methods is likely to be most
25  informative to risk assessment. Lessons learned

> **Box 3. Selection Criteria for Prototypes**
>
> - Decision context applicability (i.e., methods applicable to various types of risk management situations)
> - Data availability (i.e., both NexGen and traditional data existed to allow for validation of new approaches)
> - Illustration of a variety of methods
> - Methods
>   - ✓ Data quality
>   - ✓ Multiple, high-quality studies
>   - ✓ Consistent, coherent, and biologically plausible data
> - Active collaborations with investigators to benefit from their knowledge, modify experiments, and conduct additional analyses as needed
> - Cross-organizational collaborations fostered

26  from each prototype and group of similar prototypes will be noted as they arise and, then
27  integrated and summarized in Section 5 "Lessons Learned from Developing the Prototypes."

28  Throughout this report, characterizing systems biology is greatly emphasized. Systems biology is a
29  critical field in modern biology aimed at understanding the larger picture by integration across
30  multiple levels of biology—for example, from the gene to the molecular intermediate phenotypes
31  (e.g., gene expression), to alterations in molecular pathways and networks, and the propagation of
32  effects from cells to tissues to organs and the whole body. Systems biology also can encompass
33  subpopulation and population dynamics. Thoroughly understanding modern biology is difficult
34  without understanding systems biology. Two basic approaches are used to develop systems
35  understanding: bottom up and top down. The bottom-up approach focuses on altered molecular
36  and cellular components, and seeks to understand how the altered components fit together. This
37  approach is addressed most extensively in Tiers 1 and 2. The top-down approach focuses on larger
38  scale network interactions and disease indicators based on human clinical and epidemiologic data,
39  and associations between disease states and environmental factors (Friend 2013). This approach is
40  addressed most extensively in Tiers 3 and 2. Both the bottom-up and top-down approaches can be
41  informative, and are best used when integrated to support development of a comprehensive model.

*This document is a draft for review purposes only and does not constitute Agency policy. Do not cite or quote.*

September 2013          13

| Types of Methods and Characteristics | | Tier 1: Screening and Prioritization | Tier 2: Limited Scope Assessments | Tier 3: Major Scope Assessments |
|---|---|---|---|---|
| **Types of Methods** | **Exposure Data:** | Surrogate | Limited | Extensive environmental |
| | **Structure Information:** | QSAR Models | QSAR Models and Read Across | Mechanistic Understanding |
| | **Assay Types:** | High-throughput (HT) Assays | High Content | All informative Traditional Data |
| | **Extra Information Sources:** | Computer (*In silico*) Toxicity Models | Database Mining | Biomarkers of Exposure and Effect |
| **Characteristics** | **Time to Conduct:** | Hours–Days | Hours–Weeks | Days–Years |
| | **Cost:** | $ | $–$$ | $$–$$$ |
| | **Exposures:** | *In Vitro* | *In Vitro* and *In Vivo* | *In Vivo* |
| | **Exposure Duration:** | Short | Short | Longer |
| | **Metabolism:** | Little to None | Some | Substantial |
| | **Endpoints:** | Alterations in Key Biological Process | Alterations in Key Biological Process to Adverse Effects | Alterations in Key Biological Process, Intermediate Event to Adverse Effects/Disease |

**Increasing weight of evidence and resources to complete** →

Figure 5. Shown are the types of methods used to generate data for the prototypes and characteristics of each method. Note that all methods can be used in each decision context as available.

## 3.1. Tier 3: Major Scope Assessments

1    Tier 3 prototypes focused on chemicals with robust traditional data sets, known public health
2    outcomes, and high-confidence risk estimates. The purpose of studying these already well-
3    characterized chemicals was to better understand how new data types and methods can be most
4    effectively used in risk assessment situations where traditional data are absent or limited. In other
5    words, by "reverse engineering" from known public health risks to new types of data, it was
6    thought that potential advances in risk assessment using new types of data could be verified.
7    Molecular epidemiology, molecular clinical, and molecular *in vivo* animal data were evaluated in the
8    context of traditional information (Table 1). The Tier 3 prototypes aimed to: (1) demonstrate proof
9    of concept that new data and methods can help identify hazards and inform exposure-dose-
10   response relationships; (2) better characterize what information is most valuable for specific risk
11   assessment purposes; and (3) articulate decision considerations for identifying, analyzing, and
12   interpreting data, particularly for use in assessment of data-poor chemicals. Secondarily, this effort
13   explored how new data types can augment robust traditional data sets, and brings new insights to
14   the interpretation of traditional data.

*This document is a draft for review purposes only and does not constitute Agency policy. Do not cite or quote.*

September 2013                    14

**Table 1. Tier 3 Prototypes Approach, Including Weight of Evidence, Pros, and Cons**

| Tier 3: Major Scope Assessments<br>Environmentally-relevant *In Vivo* Exposure Studies<br>with Molecular Characterization | |
| --- | --- |
| **Approaches:** | Focuses on human data from molecular epidemiology and molecular clinical studies.<br>Includes molecularly augmented, traditional *in vivo* animal bioassays.<br>Experimentally measures dose-dependent, chemically-induced alterations in biologic functions linked to traditional intermediate events and disease outcomes.<br>Evaluates environmentally-relevant exposures.<br>Characterizes sensitive subpopulations.<br>Helps characterize impacts of various environmental factors. |
| **Weight of evidence:** | Determined by the quality and quantity of data, but can range from suggestive to known. |
| **Pros:** | Characterizes human population-associated or causal mechanisms.<br>Can inform low-dose, species-to-species and inter-individual variability, and uncertainty with data.<br>Allows extrapolation of molecular patterns to predict outcomes for less well studied chemicals. |
| **Cons:** | Are not faster or less expensive than traditional bioassays.<br>Need to control for experimental variability. |

1     The Tier 3 prototypes are benzene (and leukemia); ozone (and inflammation and lung injury); and
2     benzo[a]pyrene (B[a]P), a polycyclic aromatic hydrocarbon (PAH) (and liver cancer). The
3     prototypes focused on human data, both molecular epidemiology and molecular clinical data.
4     Human environmental exposures for the benzene and ozone prototypes were very well
5     characterized using a urinary biomarker and $^{18}O_2$ dosimetry, respectively. For B[a]P, we evaluated
6     human environmental exposures and liver cancer omics data; this evaluation was qualitatively
7     successful, but exposures were relatively poorly characterized for quantitative exposure-response
8     assessment. Hence, experimental rodent data were evaluated in addition to the human data.

9     Overall, the prototypes evaluated the use of toxicogenomics to better characterize risks, including
10    DNA transcription (transcriptomics), protein expression (proteomics), and genome-wide analyses
11    of susceptibility genes (genomics analyses of human gene variants). Some limited discussion of
12    epigenetic modification (epigenomics) in human populations is also included. Bioinformatics
13    analyses were used to evaluate toxicogenomic profiles in the context of traditional knowledge of
14    phenotypic endpoints. Each prototype:

15    •   Describes a systems biology model suitable for informing hazard identification;

16    •   Characterizes molecular biomarkers of exposure and effects suitable for characterizing
17         exposure-response at environmental concentrations;

18    •   Illustrates how multiple pathway alterations induced by environmental factors can lead to
19         and modify risks, and notes how this information might be used to characterize data-limited
20         chemicals and cumulative risks; and

1 • Identifies some gene variants that influence human susceptibility and alter risks for selected
2   subpopulations, and notes how this information could be used to characterize population
3   variability.

4 The results presented here are not intended to be a comprehensive review of all available data that
5 might be used in a risk assessment, but rather provide examples of evaluation of new data types
6 and to illustrate potential uses in risk assessment. In addition, toxicogenomics data must be
7 interpreted carefully in this context (see Text Box 4).

---

**Box 4. A Word of Caution in Interpreting Toxicogenomic Results**

Technical variability in toxicogenomic results can be a substantial source of data misinterpretation. Rigorous study design and statistical techniques increase confidence in observed associations, and increase the power to detect associations, between exposure and gene expressions particularly at low exposure levels. More generally, without such considerations, variability may obscure actual outcomes or lead to specious associations. Studies without rigorous design, data collection, and analyses are less likely to be considered appropriate for use in risk assessment.

---

8 One caveat is that the studies used in the Tier 3 prototypes were chosen mainly because they had
9 some of the most robust, concomitantly collected, traditional and new data types available. These
10 data sets demonstrate partially what can be done with new data types; however, similar data are
11 not likely to be available for many chemicals. This exercise clearly revealed that care must be given
12 to the selection of studies for new types of risk assessment, as many are insufficient for the
13 applications discussed below. The B[a]P prototype, in particular, highlights some of the challenges
14 encountered. Additionally, the fields of molecular, computational, and systems biology are in their
15 infancy in terms of application to human health risk assessment. Although results presented here
16 are promising, robust understanding and full implementation of new methods in general practice,
17 might take years, subject to the resources available for data generation and evaluation.

18 Implications for risk assessment identified by the Tier 3 prototypes are discussed at the end of this
19 section and integrated with other lessons learned in Section 5, "Lessons Learned from Developing
20 the Prototypes." It should be reiterated that the primary intention of the Tier 3 prototypes is to
21 "ground truth" approaches that could be used in more data-limited situations.

### 3.1.1. Benzene-Induced Leukemia

22 Benzene is among the 20 most widely used chemicals in the United States and is among the most
23 common environmental contaminants. A component of crude oil and gasoline, benzene is also used
24 as an intermediate in the manufacture of resins, dyes, chemical solvents, waxes, paints, glues,
25 plastics, and synthetic rubbers. The major sources of benzene exposure are anthropogenic and
26 include fixed industrial sources, fuel evaporation from gasoline filling stations, and automobile
27 exhaust. Benzene has been measured in outdoor air at various locations in the United States at
28 concentrations ranging from 0.02 ppb (0.06 $\mu g/m^3$) in a rural area to 112 ppb (356 $\mu g/m^3$) in an
29 urban area (IARC 2012). Personal monitoring of benzene exposure in Detroit, Michigan, reported a
30 mean of 1.72 ppb (5.5 $\mu g/m^3$) (George et al. 2011). The maximum contaminant level (MCL) in
31 drinking water is 5.0 $\mu g/L$ or 5 ppb (EPA 2012b). The OSHA permissible exposure limit (PEL) for

---

1 benzene workers in the United States is 1 ppm
2 (https://www.osha.gov/dts/chemicalsampling/data/CH_220100.html).

3 Benzene is a known human carcinogen (IARC 2012, ATSDR 2007, EPA 2000, NIOSH 1992).
4 Epidemiologic studies have shown that benzene exposure leads to an increased risk of acute
5 myeloid leukemia (AML), myelodysplastic syndrome (MDS), hematotoxicity (toxicity to the blood),
6 and other blood disorders (IARC 2012, Schnatter et al. 2012, EPA 2000, Goldstein 1988). AML is
7 characterized by uncontrolled proliferation of clonal neoplastic cells and accumulation in the bone
8 marrow, with an impaired differentiation program. AML accounts for about 30% of all adult
9 leukemias and is the most common cause of leukemia death (Howlader et al. 2013). Studies also
10 indicate that benzene might cause lymphoma and childhood leukemia (Smith, MT et al. 2011).

11 The extensive molecular epidemiologic and molecular clinical data sets available for both benzene
12 and leukemia are ideal to explore how new data types might be used to inform risk assessments.
13 The work described here focuses on studies where traditional and molecular data were collected
14 simultaneously using a variety of omic methods, including genome-wide analyses of susceptibility
15 genes (using genomic methods), protein expression (proteomics), and epigenetic modification
16 (epigenomics) (McHale et al., 2012). The studies also were conducted over a range of
17 environmental exposure levels (<0.1 ppm to ≤ 10 ppm). The information was developed primarily
18 by Martyn Smith and colleagues (University of California, Berkeley). Systems biology of benzene-
19 induced leukemia is summarized in McHale et al. (2011) and Smith et al. (2011).

## Systems Biology of Benzene-Induced Disease

20 Although benzene is among the most well-studied environmental chemicals, understanding the
21 molecular mechanisms underlying hematopoietic cancer is somewhat recent (see Text Box 5 for a
22 brief description). In 2009, McHale et al. (2012) identified exposure-dependent alterations in genes
23 and pathways (in peripheral blood mononuclear cells using transcriptomics), and hematotoxicity
24 associated with benzene exposure
25 (>10 ppm) in occupationally
26 exposed Chinese workers. McHale
27 et al. (2011) extended these
28 findings to lower exposure levels
29 (<1 ppm to ≤ 10 ppm). (The
30 current U.S. occupational
31 standard is 1 ppm.) In subsequent
32 work, Thomas et al. demonstrated
33 changes in gene expression at
34 current U.S. urban levels in
35 Chinese workers exposed to levels
36 <0.1 ppm. The exposure-response
37 models used in these analyses
38 were not selected *a priori*, but
39 rather driven by the best fit of the
40 data. Results are consistent with
41 supralinear exposure-responses,
42 which have also been reported in
43 traditional epidemiology studies (Lan et al. 2004).

> **Box 5. Molecular Mechanism of Acute Myeloid Leukemia (AML)**
>
> The probable mechanism by which benzene induces leukemia involves the "targeting of critical genes and pathways" (McHale et al. 2012). Benzene has the potential to induce abnormalities in the genes, chromosomes, or epigenetic mechanisms of hematopoietic stem cells (HSC). It can also disrupt its normal cell cycle, leading to apoptosis, increased cell proliferation, and altered differentiation of the HSCs. Benzene causes these effects and ultimately leukemia through oxidative stress, dysregulating proteins that control normal functioning of HSCs, and reducing the ability of the body to detect and destroy cancerous cells (McHale et al. 2012).
>
> For AML specifically, two events that are important for leukemic transformation have been identified. The first event is uncontrolled cell growth, which is mediated by upregulation of cell survival genes. The second event is alteration of transcription factors that control the HSC differentiation. That is, the transcription factor proteins can be mutated or can target certain genes in a way that interferes with the appropriate differentiation of HSCs (Kanehisa Laboratories 2013, Wang, I et al. 2012).

1    The systems biology of benzene-induced early effects have been articulated by McHale et al. (2012)
2    and others (Smith, MT et al. 2011, Zhang, L et al. 2010). Benzene-induced leukemia is thought to be
3    initiated when metabolites of benzene target genes or pathways that are critical to hematopoiesis
4    in hematopoietic stem cells. Interactions among various cell types within the bone marrow and
5    among various tissues also play a role in leukemia (e.g., immunosurveillance). The underlying
6    mechanisms of benzene-induced leukemia, shown in Figure 6, center on exposure-dependent
7    pathway alterations comprising 147 significantly altered genes (cross validated on two microarray
8    test platforms [Illumina and Affymetrix]). The gene expression profile changes with dose, with
9    some genes (and related biological processes) being expressed at all levels, while others are
10   expressed only at higher concentrations. Of the 147 genes, the expression of 16 genes was
11   significantly altered at all exposure levels. These 16 signature genes are involved in immune
12   response, inflammatory response, cell adhesion, cell matrix adhesion, and blood coagulation, and
13   are most strongly associated with AML disease pathways (McHale et al., 2011). This set of 16 genes
14   forms a biomarker for exposure (and associated leukemia) for future work, particularly in
15   augmenting traditional epidemiology studies and enabling new types of molecular epidemiology
16   studies at lower concentrations. As will be discussed later in this section, understanding of the
17   systems biology and molecular initiating events (MIEs) in leukemia can also potentially enable
18   screening of relatively unstudied chemicals for similar signature events. Clinical studies of
19   chemotherapeutic agents, which alter gene expression in these same pathways and are used in the
20   treatment of leukemia add evidence to the causal relationships between specific gene/pathway
21   alterations and leukemia (Hatzimichael and Crook 2013).

22   In addition to leukemia, a lymphoma disease signature is evident with benzene exposure (McHale
23   et al. 2012, McHale et al. 2011, Smith, MT et al. 2011). The traditional epidemiology data on
24   lymphoma are not conclusive. Characterization of a benzene-induced molecular mechanism for
25   lymphoma adds considerably to the weight of evidence for benzene-induced lymphoma,
26   highlighting the use of molecular mechanism or mode-of-action information to strengthen weight-
27   of-evidence determinations (IARC 2012).

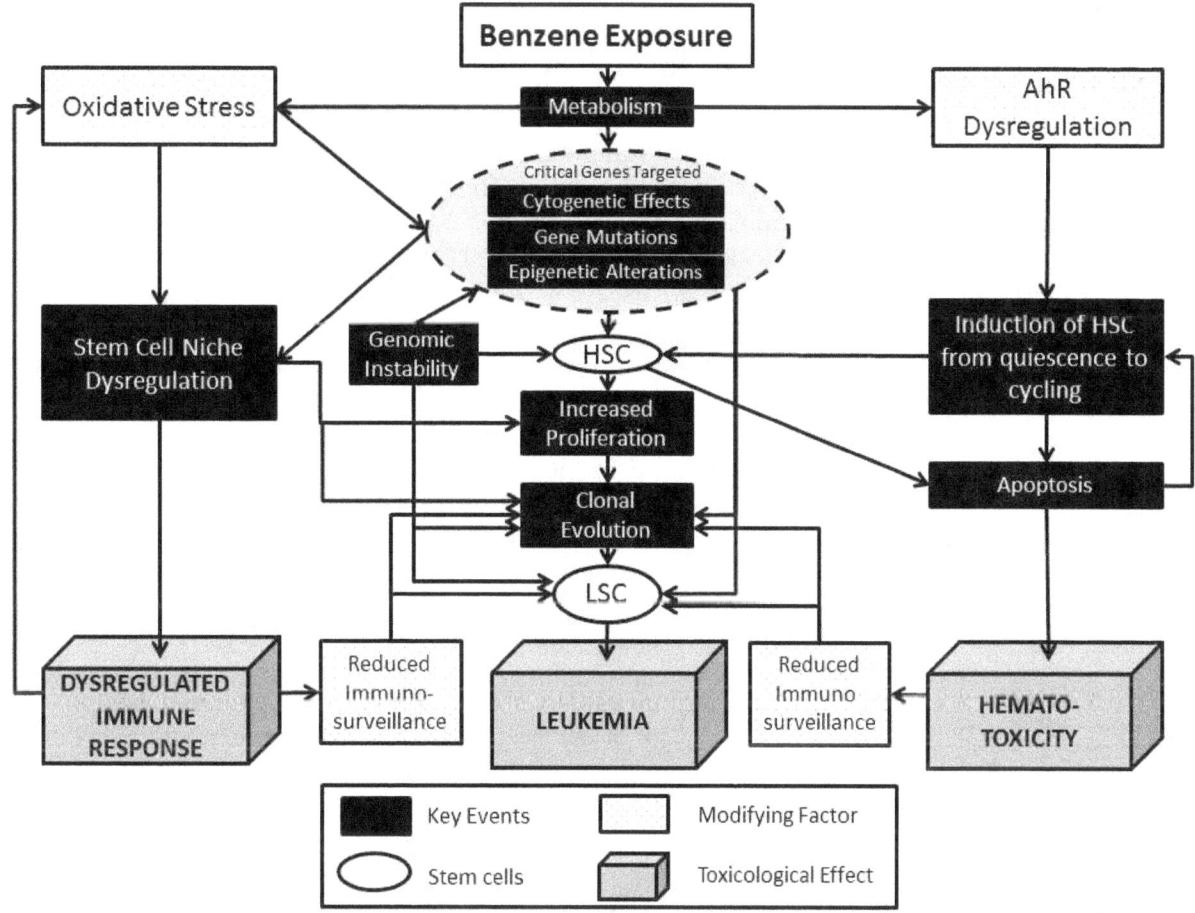

Figure 6. Multiple modes of action (MOAs) of benzene-induced leukemogenesis. Potential key events, modifying factors, and toxicological effects are depicted in the legend. Stem cells can be either HSCs (hematopoietic stem cells) or LSCs (leukemic stem cells) (Smith, MT et al. 2011), reproduced with permission from Elsevier.

## De Novo and Other Chemical Leukemogen-Induced Disease

1    Interestingly, molecular mechanisms for benzene-induced leukemia appear similar to *de novo*
2    (without an obvious cause) AML and AML induced by other environmental agents (e.g., alkylating
3    agents, topoisomerase II inhibitors) (IARC 2012, McHale et al. 2012, Pedersen-Bjergaard et al.
4    2008). Figure 7a[6] shows a network of genes and pathways involved in *de novo* and chemically
5    induced leukemia [Kyoto Encyclopedia of Genes and Genomes (KEGG) (Kanehisa Laboratories
6    2013)]. The circles in the figure indicate some of the specific genes and pathways affected by
7    leukemogenic agents and environmental modifiers (Kanehisa Laboratories 2013, IARC 2012,
8    McHale et al. 2011, Pedersen-Bjergaard et al. 2008). Additional evidence for the causal role for
9    these genes and pathways in AML is provided by the study of human genetic variants associated
10   with altered risks and chemotherapeutics that reverse adverse alterations in some of these same

---

[6]The basic AML network figure used in Figures 7a and 7b is from the Kyoto Encyclopedia of Genes and Genomes (KEGG) (Kanehisa Laboratories 2013)). The added circles are the work of the report authors.

1  genes and pathways (discussed below). Although mechanistically similar, different agents can
2  display specific characteristics; including origins in cells at different stages of hematopoiesis,
3  distinct cytogenetic subtypes, and different latencies (Irons et al. 2013, McHale et al. 2012).
4  Figure 7a highlights how a disease network can be modified at different points but still lead to a
5  common disease outcome. These mechanistic commonalities and differences among *de novo* and
6  chemically-induced health effects can be used to characterize chemicals with limited data, which
7  have nevertheless been shown to induce mutations, chromosome changes, or specific changes in
8  gene expression. Data-limited chemicals would be of elevated concern if they are shown to alter
9  pathways similar to that observed in *de novo* disease or with well-studied leukemogens. For
10  example, see the work in Thomas R et al. (2012) where the authors used existing information on
11  gene and protein targets of 29 known leukemia-causing chemicals and 11 carcinogens that are not
12  known to cause leukemia, the authors were able to develop a classification scheme that could
13  distinguish a random leukemia-causing/nonleukemia-causing carcinogen pair with a 76%
14  probability. Provided later in this section (in the ozone and B[a]P prototypes) is similar evidence
15  for the importance of networks when considering chemical-related diseases, similarities of
16  chemical-related and *de novo* diseases, and the role of mechanisms in improved understanding of
17  cumulative risks.

## Cumulative Risks from Environmental Factors

18  Evidence suggests that, in addition to environmental exposures, genetic variations and lifestyle
19  factors such as smoking, obesity, diet, and alcohol use are risk factors for leukemia (Smith, MT et al.
20  2011, Pedersen-Bjergaard et al. 2008, Belson et al. 2007, Ilhan et al. 2006). Environmental
21  exposures of the developing organism could also be a risk factor for disease later in life, given the
22  potential of benzene and other environmental agents to alter epigenetics, the sensitivity of the
23  developing organism to epigenomic changes, and the association of environmental exposures and
24  childhood leukemias (Boekelheide et al. 2012). Figure 7a shows how multiple environmental
25  factors can alter several molecular events in a manner that alters risks, and how mechanistic
26  knowledge might be used to identify or exclude chemicals based on common mechanisms and
27  impacts on cumulative risks.

28  Individuals exposed to known environmental and lifestyle risk factors are estimated to account for
29  approximately 20% of acute leukemia incidences, indicating that host genetic susceptibility might
30  be instrumental in the development of leukemia (Smith, MT et al. 2011). By identifying mechanistic
31  commonalities, or the lack thereof, among chemicals, new omic approaches can provide tools for
32  characterizing roles that intrinsic and extrinsic risk factors might play in individual and
33  subpopulation risks. Below we discuss genetic variation more specifically and provide an example
34  of altered subpopulation risks based on genetic variations.[7]

## Genetic Variation and Susceptibility in the Human Population

35  Genetic susceptibility for developing AML, and how it relates to chemical risks, has been studied by
36  several investigators (Zhuo et al. 2012, North et al. 2011, Shen et al. 2011, Smith, MT et al. 2011,
37  Garte et al. 2008). Several genetic variations in individual genes appear to increase risks for

---

[7]Human genetic variation is the genetic differences among subpopulations. Multiple variants of any given gene might occur in the population. These differing DNA codings determine distinct traits or polymorphisms that can influence risks.

1    developing AML, while at least one decreases risks. Sillé et al. (2012) reported 12 independent risk
2    loci (specific regions within the genome, which can be a single base, as in this case, or an entire
3    gene) with the potential to alter gene expression. A significant number of variants (single
4    nucleotide polymorphisms [SNPs] or single nucleotide variations) related to a tumor suppressor
5    gene, signaling pathways, or residing in putative regulatory elements,[8] have been linked to various
6    types of multiple hematological cancers. Figure 7b highlights genes that vary in the human
7    population and are associated with altered leukemia risks (Hatzimichael and Crook 2013, Kanehisa
8    Laboratories 2013). Figure 8 provides an example of differential risks resulting from one human
9    variant.[9] The overall data indicated a significant variation in risk (42%) relative to the CYP1A1
10   genotype (Zhuo et al. 2012). The shift in odds ratio is also shown in Figure 8.

11   When one considers that many genes are associated with benzene-induced leukemia, the potential
12   for variation in subpopulation risks via individual genes, combinations of genes, and gene variants
13   becomes apparent. Other risk factors (e.g., lifestyle) would add to the human variability in response.
14   As discussed in Section 4, NexGen approaches exist that can facilitate characterization of human
15   variability as never before.

---

[8]Putative regulatory elements are areas of the gene that do not code for proteins but rather regulate DNA transcription into proteins.
[9]SNP leads to a base substitution of isoleucine with valine at codon 462 in exon7 (Ile462Val or CYP1A1*2C polymorphism, rs1048943). Thus, the exon7 restriction site polymorphism results in three genotypes: a predominant homozygous Ile/Ile, the heterozygote Ile/Val, and a rare homozygous Val/Val.

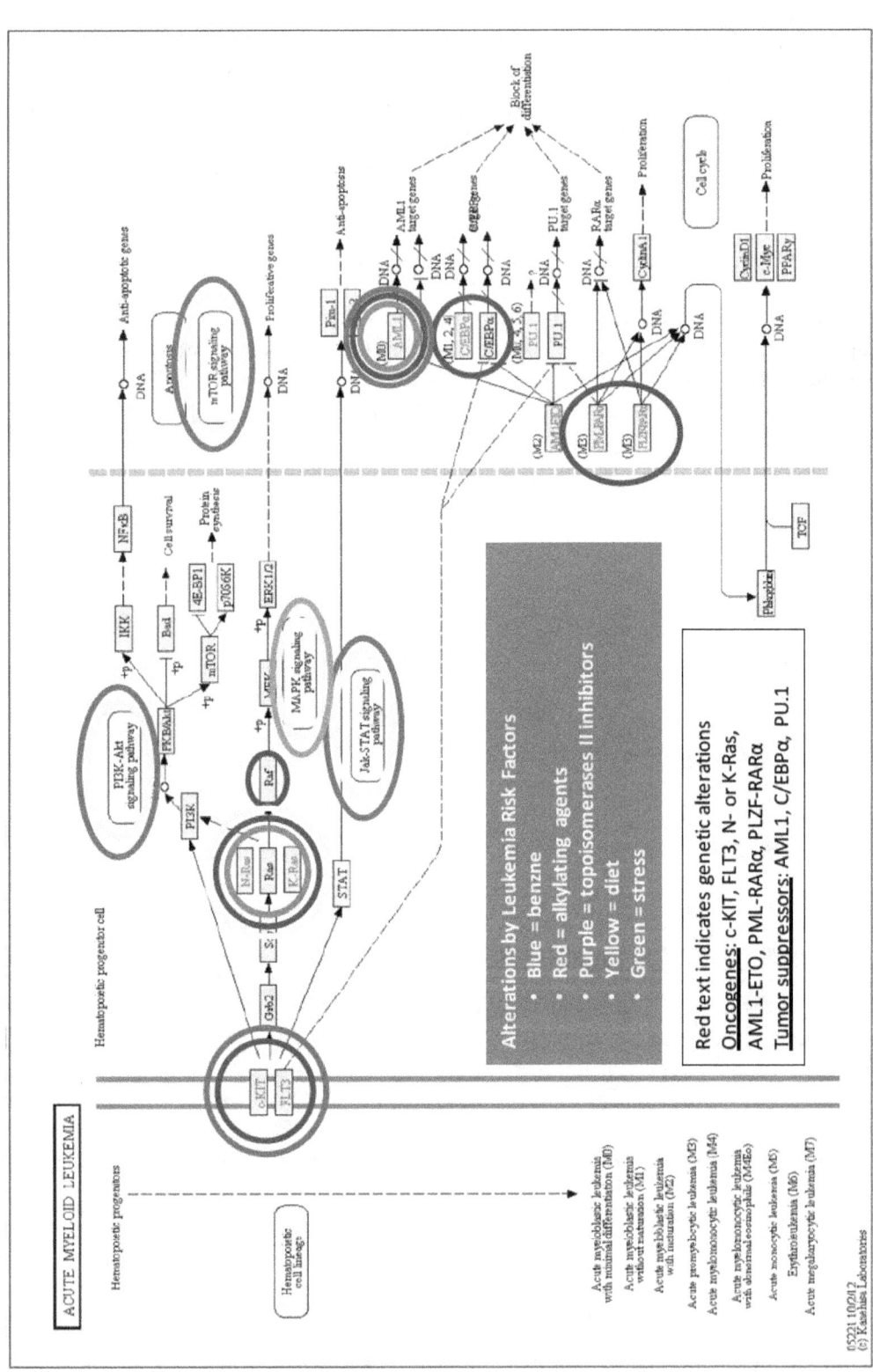

Figure 7a. The Kyoto Encyclopedia of Genes and Genomes (KEGG) diagram (http://www.genome.jp/kegg-bin/show_pathway?hsa05221) illustrates some of the currently understood molecular pathways involved in acute myeloid leukemia (AML). Altered oncogenes and tumor suppressor genes are noted in red type (Kanehisa Laboratories 2013). The circles (added by authors) note specific genes and pathways that are modified by benzene, other chemical leukemogens, and other risk factors. While intended to be illustrative rather than comprehensive, it can be seen how single or combinations of environmental factors could modify risks for leukemia, and how such knowledge could be used to evaluated joint effects of environmental factors (IARC 2012, McHale et al. 2012, Smith, MT et al. 2011, Pdersen-Bjergaard et al. 2008).

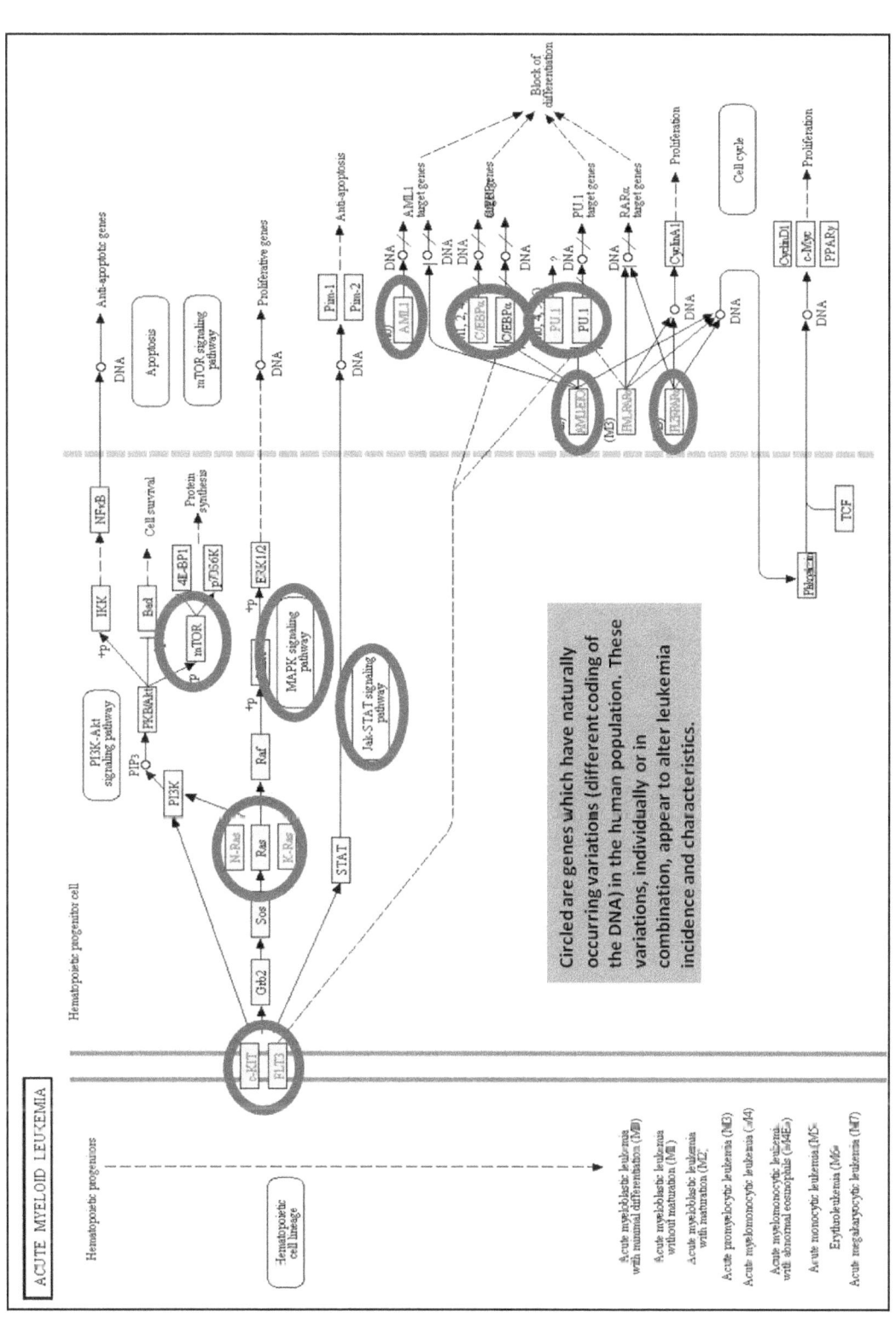

Figure 7b. This figure shows the same acute myeloid leukemia (AML) KEGG diagram (Kanehisa Laboratories 2013) as shown in Figure 7a, with circles added by authors. In this version, circled are the locations of naturally occurring human genomic variants that increase the risk of AML (Hatzimichael and Crook 2013, Sille et al. 2012). Characterizing genomic variant subpopulations and associated risks can help us to better describe human variability and susceptibility for specific diseases.

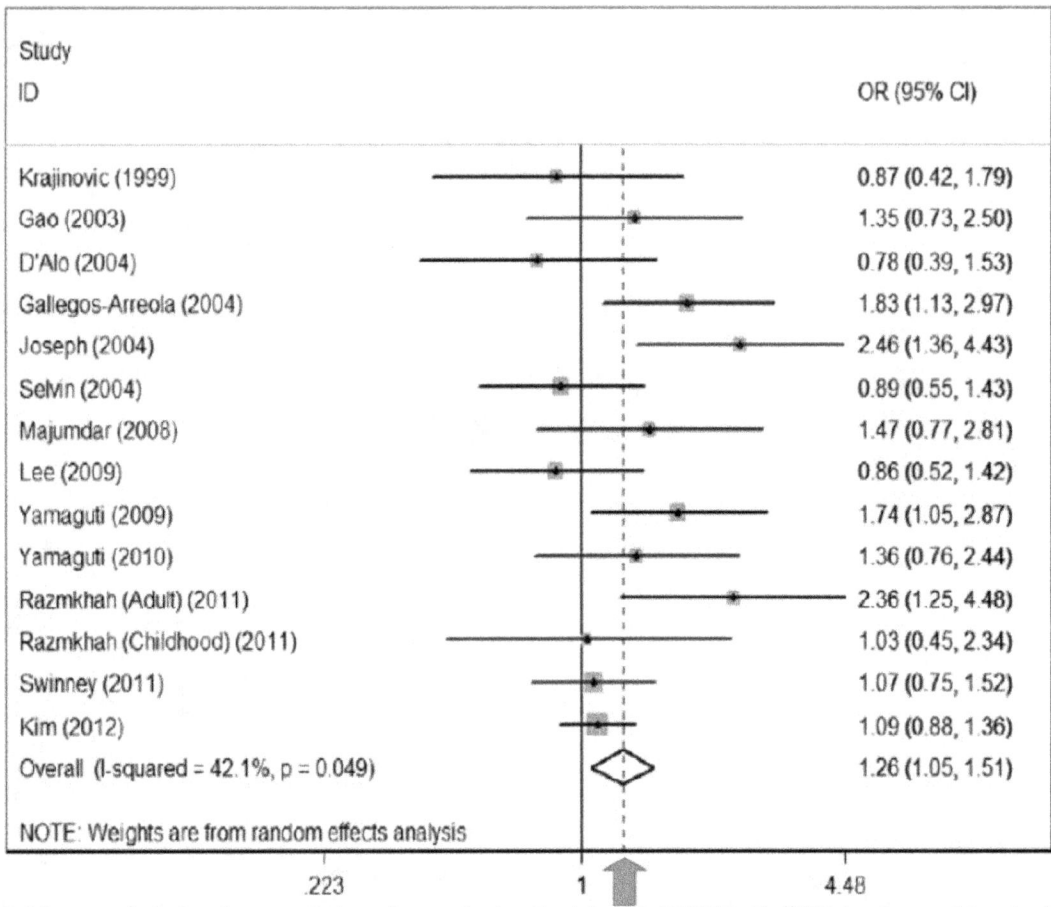

| Study ID | | OR (95% CI) |
|---|---|---|
| Krajinovic (1999) | | 0.87 (0.42, 1.79) |
| Gao (2003) | | 1.35 (0.73, 2.50) |
| D'Alo (2004) | | 0.78 (0.39, 1.53) |
| Gallegos-Arreola (2004) | | 1.83 (1.13, 2.97) |
| Joseph (2004) | | 2.46 (1.36, 4.43) |
| Selvin (2004) | | 0.89 (0.55, 1.43) |
| Majumdar (2008) | | 1.47 (0.77, 2.81) |
| Lee (2009) | | 0.86 (0.52, 1.42) |
| Yamaguti (2009) | | 1.74 (1.05, 2.87) |
| Yamaguti (2010) | | 1.36 (0.76, 2.44) |
| Razmkhah (Adult) (2011) | | 2.36 (1.25, 4.48) |
| Razmkhah (Childhood) (2011) | | 1.03 (0.45, 2.34) |
| Swinney (2011) | | 1.07 (0.75, 1.52) |
| Kim (2012) | | 1.09 (0.88, 1.36) |
| Overall (I-squared = 42.1%, p = 0.049) | | 1.26 (1.05, 1.51) |

NOTE: Weights are from random effects analysis

.223     1     4.48

Figure 8. Meta-analysis for the association of acute leukemia risk with CYP1A1. Ile462Val polymorphism is shown (OR = odds ratio). The overall risk was 42% greater (95% CI = 1.11–1.98) for Val/Val+Val/Ile versus Ile/Ile (Zhuo et al. 2012). Reproduced with permission from PLoS One.

### *In Vitro* Evaluation of Toxicogenomic Signatures

1  As has been previously noted, the primary function of the Tier 3 prototypes is to inform how we
2  evaluate data-limited chemicals. Hence, a comparison of *in vivo* and *in vitro* benzene results is
3  discussed here. Godderis et al. (2012) conducted an *in vitro* study in TK6 cells to detect gene
4  signatures and biological pathway perturbations, using global gene expression analysis, resulting
5  from exposure to 15 genotoxic carcinogens, including benzene and its metabolites. The goal was to
6  determine if well-characterized chemicals could be used to characterize data-limited chemicals by
7  comparing gene signatures. Although pathways altered by exposure to benzene and its metabolites
8  were in general agreement with previous *in vivo* studies, the authors pointed out that several
9  factors can complicate comparison of *in vivo* and *in vitro* data, for example, metabolism and cell
10  types. The authors concluded that use of toxicogenomic signatures hold great promise for
11  evaluation of data-limited chemicals. They noted that for the carcinogens in the study, some *in vitro*
12  processes mapped against known or likely carcinogenic processes, but determining discriminatory
13  mechanisms based on *in vitro* data alone was difficult. This observation suggests that the approach
14  of developing putative mechanisms of action based on data-rich meta-analyses of human disease

*This document is a draft for review purposes only and does not constitute Agency policy. Do not cite or quote.*

September 2013     24

1   and mapping *in vitro* data against these models might prove more successful than attempting to
2   understand mechanisms of action based on *in vitro* data alone.

## Risk Assessment Implications from the Benzene Prototype

3   The benzene prototype demonstrated how molecular biology data, particularly mechanistic
4   signatures, can be used in hazard identification and exposure-dose-response assessment.

5   *Hazard Identification* – Specifically, genes and pathways altered by benzene exposures are
6   strongly associated with a network of pathways associated with known (AML) and likely
7   (lymphoma) outcomes. Additional evidence for a causal relationship between alterations in specific
8   genes and pathways and increased leukemia risk is provided by observed similarities in pathway
9   disruptions: (1) caused by other chemical leukemogens, (2) observed in leukemia of unknown
10  origins, and (3) reversed by certain leukemia chemotherapeutic agents. Hence, observations from
11  both molecular epidemiology and molecular clinical studies provide evidence that molecular
12  signatures can predict specific diseases with some confidence. These data suggest that well-defined
13  pathway and network disruptions strongly associated with a specific disease could be used to
14  screen chemicals with limited molecular data for their potential to increase risks for the specified
15  disease by causing similar mechanistic disruptions. Anchoring of the molecular patterns to apical
16  outcomes, considerable systems biology knowledge, and high-quality data; however, appear
17  necessary to define the disease signature against which data-limited chemicals could be compared.

18  *Exposure-Dose-Response Assessment* – A specific exposure-dose-dependent gene signature for
19  leukemia was observed at all environmental exposure concentrations measured (<0.1 to >10 ppm);
20  the magnitude of signature expression varied in a dose-dependent manner. This signature is a
21  biomarker of both exposure and effect. Such signatures or biomarkers can extend the exposure
22  range of traditional epidemiologic studies to lower exposures and reduce measurement error. This
23  type of data can measure low-dose-response relationships and, potentially, mitigate a source of
24  substantial controversy in chemical risk assessment, that is, low-dose extrapolation. In the future,
25  one can envision routine replacement of low-dose extrapolation with measurements of molecular
26  signatures. The established dose-response for specific gene signatures could be used to estimate
27  the potency or relative potency for data-limited chemicals. In particular, ranking of chemicals is
28  feasible when using similar protocols such as those characteristic of Toxicology in the 21st Century
29  (Tox21) or ToxCast™.

30  The exposure-response models used in this prototype were not specified in advance, but the choice
31  relied on the best fit from among multiple models. Hence, the model was "agnostic" on the issues of
32  threshold/no threshold and the shape of the low-exposure-response relationship. Such an approach
33  would mitigate another source of controversy in risk assessment, that of model choice.

34  *Cumulative Risk Assessments* – Understanding of a common mechanism of action for multiple
35  environmental factors can allow for improved cumulative risk assessments. It should be noted that
36  overly simplified descriptions of mode of action (MOA) or adverse outcome pathways (AOPs) could
37  miss interactions among the environmental factors as shown in Figure 7a.

38  *Variability and Susceptibility in Human Response* – An example of risk characterization
39  associated with different genetic variations is provided. With additional research and data evolving
40  from personalized medicine, the understanding of population variation and distribution of

1    responses in the human population could be improved. These data also could help improve
2    estimates of the size of sensitive subpopulations.

### 3.1.2.  Ozone-induced Lung Inflammation and Injury

#### Use of Ozone as a Model Pollutant

3    Hundreds of controlled human exposure studies have described biological changes in volunteers
4    exposed acutely (usually for 2–6 hours) to concentrations of ozone ranging from 0.06 to 0.4 ppm
5    (EPA 2011d).[10] These studies show that exposure to ozone results in decrements in several indices
6    of lung function, increases in markers of pulmonary inflammation, and alterations in host defenses
7    against inhaled pathogens and lung injury. This database represents the single largest human
8    database of any pollutant EPA has studied. As a consequence and because the mechanisms are well
9    understood, the database provides an ideal opportunity to demonstrate proof of concept for use of
10   molecular biology data to inform assessment of human risks, to develop decision considerations for
11   use of such data, and to explore the value of various types of information.

12   The underpinning of an AOP-based paradigm in risk assessment methodology is the concept of
13   studying biological pathways. The perturbation of a biological pathway initiates a set of key events
14   that cause an adverse outcome associated with an environmental stressor. If such pathway
15   responses are known and represented by a set of quantitative *in vitro* assays, the results of these
16   assays can be used to build quantitative biological activity relationships. Coupling these results with
17   appropriate physiologically based pharmacokinetic (PBPK) modeling and exposure estimates for
18   estimating tissue doses can be useful for hazard identification and dose-response assessment. For
19   *in vitro* pathway information to be used in risk assessment, the quantitative relationship between
20   perturbation of a pathway following *in vitro* exposure
21   and downstream endpoints (i.e., pathophysiological
22   changes at the tissue or organism level following *in vivo*
23   exposure of animals or preferably humans) must be
24   established. This framework is not likely to be possible,
25   however, as sufficient *in vivo* data are lacking for most of
26   the toxicants that EPA is responsible for regulating
27   (Crump et al. 2010). Therefore, using model systems in
28   which both *in vitro* and *in vivo* data are available is
29   necessary to validate how well pathway information
30   from the former can predict human responses to
31   toxicants. Ozone provides such a model system for lung
32   inflammation and injury (see Text Box 6 for a
33   description of inflammation). This model can be used
34   for less well-studied chemicals to identify and
35   characterize their potential to induce lung inflammation
36   and injury. Figure 9 outlines physiological and cellular pathways by which ozone causes
37   pathophysiological changes in humans via the lung response. This prototype focuses on pathways
38   that lead to inflammation, which are shown in the open boxes. Several human studies characterize
39   inflammation at multiple ozone concentrations during and after exposure, providing a rich data set

> **Box 6. Inflammation**
> Inflammation is the immune system's response to damage to cells and organs by pathogens, chemicals, or physical insult. Initially, inflammation involves changes in local blood flow and accumulation of various inflammatory cells (e.g., neutrophils, lymphocytes) at the site of injury. Pathogens and cell debris caused by the inflammatory response are then removed as tissues begin to repair. If the delicate balance between inflammation and resolution of the events leading to the inflammation is dysregulated, or tissue insult continues, inflammation can lead to disease pathology (Wang, I et al. 2012, Medzhitov 2008).

---

[10]The current ozone standard calls for limitation of the fourth highest daily maximal 8-hour ozone concentration in a year to 0.075 ppm, based on a 3-year average.

1    of human *in vivo* responses. Additional pathways based on neurological responses to ozone
2    exposure that are not captured in this figure also might be possible.

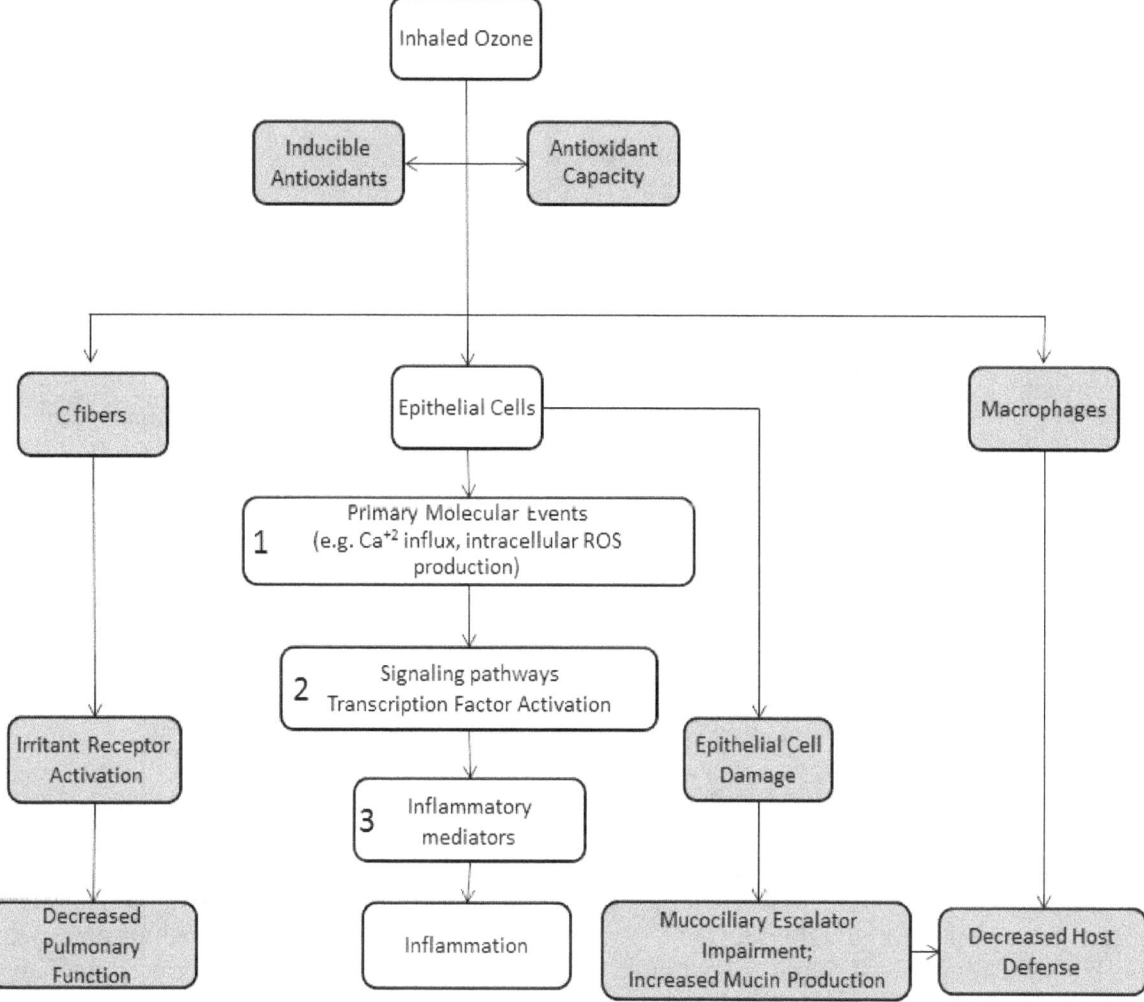

Figure 9. Framework diagram of ozone key events and modes of action (MOAs) related to lung injury occurring *in vivo*.

## Challenges with Using an AOP Approach for Risk Assessment

3    Using model systems based on *in vitro* pathway information to predict human *in vivo* responses to
4    toxicants for risk assessment purposes presents certain challenges. A major hurdle relates to
5    extrapolation from *in vitro* to *in vivo* effects. Many *in vitro* approaches use animal cells or
6    transformed cell lines derived from humans, which might not accurately reflect cell interactions or
7    events in the pathway for human *in vivo* effects. For example, a parent toxicant might be biologically
8    transformed into a more active form by cells that are not represented in the *in vitro* system (e.g.,
9    liver cells) before interacting with the target cells represented in the assay. In the lungs, epithelial
10   cells that line the human airways are the first and primary targets of inhaled toxicants and are
11   believed to be the cells that initiate lung inflammation. Studies have shown that pathways in the

1    cultured *in vitro* cells that have been activated by air pollutants are also altered in these same cells
2    following *in vivo* exposure to the same pollutant (Selgrade et al. 1995). This ability to show
3    concordance between *in vitro* and *in vivo* exposures thus is a major advantage of the modeled lung
4    system discussed here.

5    A second challenge associated with *in vitro* approaches is ensuring that the dose of toxicant
6    delivered to cultured cells is similar to that which these cells would encounter following an *in vivo*
7    exposure. Frequently, cultured cells are exposed to toxicant levels that are orders of magnitude
8    greater than they would be *in vivo*. There is no assurance that the same biological pathways are
9    adversely affected in both situations. Ozone, however, can be prepared using the heavy oxygen
10   isotope ($^{18}O_2$), which can be separated from $^{16}O_2$ and quantified by mass spectroscopy. When ozone
11   attacks a target tissue, the $^{18}O_2$ tag is bound to that tissue. This approach has been used to
12   normalize the dose of ozone delivered to rats and humans (Hatch et al. 1994) and to support
13   estimates that target tissue doses in rats exposed to 2.0 ppm ozone are comparable to target tissue
14   doses in humans exposed to 0.4 ppm ozone. This same approach can be used to normalize the dose
15   of ozone delivered to cultured cells and humans.

16   Ozone is one of the few pollutants for which an extensive animal and human health effects database
17   is available. Coupled with *in vitro* pathway data, this prototype pollutant can be used to illustrate
18   both how a biologically based dose-response modeling approach can be used to provide this
19   framework and how a systems biology model and genomics data can be used for risk assessment.

## AOP Studies

20   *In Vivo Studies* – Young, healthy volunteers were exposed to filtered air and a relevant
21   concentration of ozone (0.30 ppm) previously shown to induce a measurable inflammatory
22   response. Bronchoscopy was used to obtain cells and lung fluid at 1 and 24 hours after exposure. To
23   ensure that pathophysiological effects observed in this study were comparable to those reported in
24   earlier studies, downstream biomarkers of inflammation such as the influx of neutrophils were
25   measured (Devlin et al. 2012), as were markers of cell injury (lactate dehydrogenase) and leakage
26   of plasma components across the damaged epithelial cell barrier (albumin) into the lung airways.
27   Bronchial airway epithelial cells were obtained by brush scraping, and the microarray technology
28   was used to define pathways affected by *in vivo* ozone exposure. In addition, quantitative
29   proteomics was used to correlate changes in messenger ribonucleic acid (mRNA) measured by
30   microarray with changes in their protein counterparts (see Figure 9, event 3).

31   *In Vitro Studies* – A subset of airway epithelial cells was collected from volunteers following
32   exposure to filtered air and cultured at an air-liquid interface. These cells were exposed to
33   concentrations of ozone that had been shown (from the results of $^{18}O_3$ experiments) to be
34   comparable to the dose of ozone encountered by airway epithelial cells following a specified *in vivo*
35   exposure. This approach allows comparison of an *in vitro* and *in vivo* response of cells from the
36   same person for comparable exposures. Similar to the *in vivo* studies, microarray and proteomics
37   were used to identify and define pathways affected by ozone in these cells.

38   *Signaling Pathways* – Upstream signaling events shown in Figure 9, event 2 (e.g., transcription
39   factor activation, MAP kinase pathways, production of reactive oxygen species [ROS]) was assessed
40   to determine the MOA by which ozone activates downstream batteries of pro-inflammatory genes.
41   Pathways that are altered by exposure of cultured airway epithelial cells to ozone can be compared
42   with those altered in airway epithelial cells of the same person exposed *in vivo* to ozone. A
43   comparison can be made to determine the accuracy of the *in vitro* system in mimicking events

1  following exposure in the *in vivo* system and to assess differences in the variability of the response.
2  Figure 10 illustrates potential upstream signaling pathways that could be induced by ozone and
3  lead to activation of downstream batteries of pro-inflammatory genes.

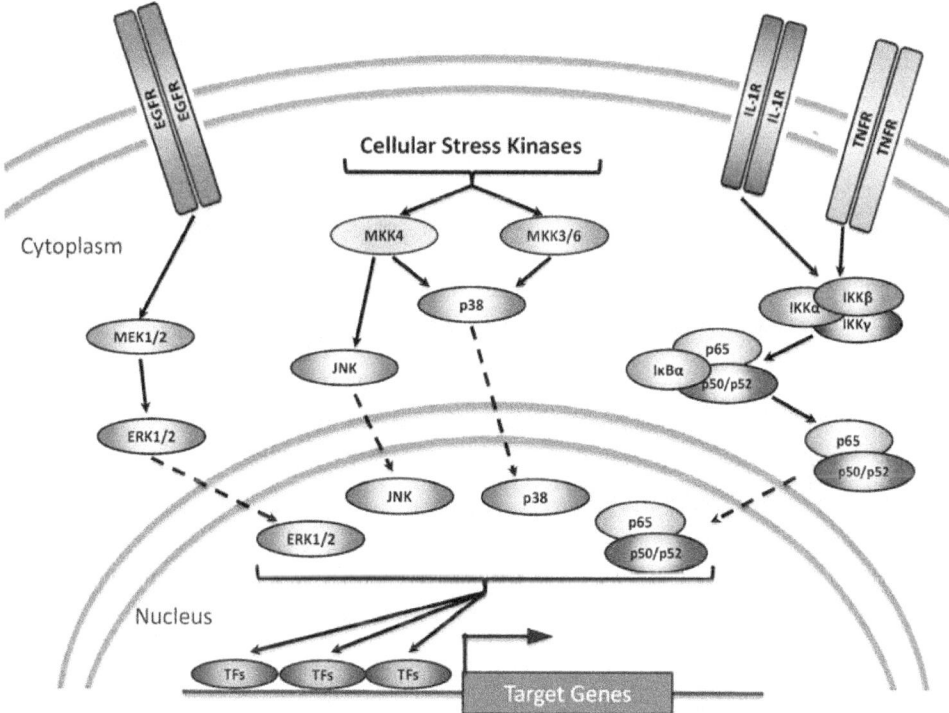

Figure 10. Potential pathways by which ozone causes production of pro-inflammatory mediators in
epithelial cells.

4   Microarray technology was used to determine which of these pathways is most likely to be altered
5   by ozone exposure. The two most highly scoring molecular networks following exposure of
6   cultured airway epithelial cells to ozone *in vitro* are presented in Figure 11. The networks involve
7   modulation of genes in NF-κB and extracellular signal-regulated kinase signaling pathways. The
8   gene list input to the Ingenuity Pathway Analysis was generated by combining all genes found to be
9   differentially expressed immediately following a 2-hour exposure of bronchial epithelial cells to
10  0.25, 0.50, 0.75, and 1.0 ppm ozone or clean air. Exposure-dose was normalized using $^{18}O_3$
11  dosimetry from *in vitro* and *in vivo* human studies. Networks are displayed with representative
12  symbols for the protein products of the mRNA transcripts. Red represents putative upregulated
13  transcripts induced by ozone, and green represents putative downregulated transcripts in response
14  to ozone. Additional molecules from the Ingenuity Knowledge Base, which were not present in the
15  differentially expressed gene (DEG) list, are uncolored in the networks. The same putative
16  networks were also identified in epithelial cells removed from human airways 1 hour after *in vivo*
17  exposure to ozone.

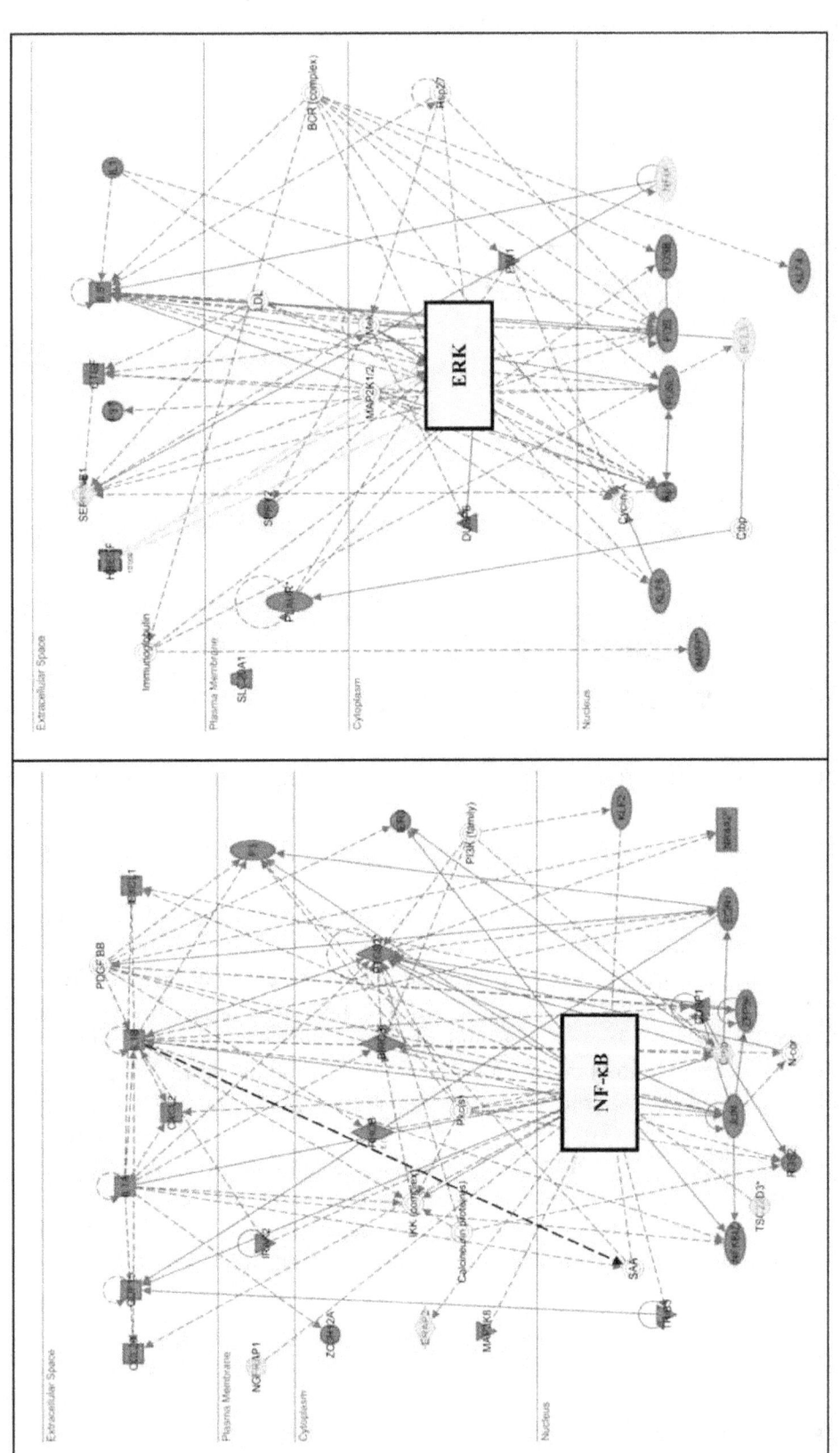

Figure 11. Molecular pathway analysis by Ingenuity Pathway Analysis.

## Primary Molecular Events

1   Many pollutants induce intracellular oxidative stress, which can affect signaling pathways and
2   ultimately lead to activation of batteries of pro-inflammatory genes. One pathway by which this
3   might occur (Figure 12) is activated in cultured human airway epithelial cells exposed to
4   particulate air pollution. Ozone is an inherently
5   potent oxidant and is known to cause oxidative
6   damage to lipids, proteins, and nucleic acids.
7   Until recently, whether ozone also induced
8   intracellular ROS was unknown. Figure 13
9   shows that ozone can induce a rapid dose- and
10  time-dependent increase in cytosolic
11  intracellular glutathione redox potential, a
12  measure of ROS (Gibbs-Flournoy et al. 2013).
13  Whether the ROS produced following ozone
14  exposure actually activates downstream
15  signaling pathways via the mechanism shown in
16  Figure 12 is unknown.

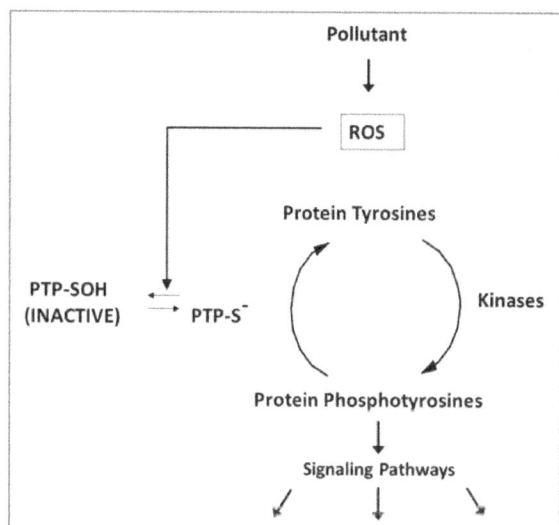

Figure 12. Role of reactive oxygen species (ROS) in mediating pollutant-induced inflammation.

## System Biology Modeling

17  Quantitative systems biology models are
18  translational, and their development is data
19  driven, with model structure and dynamics
20  parameterized using data on (1) basic biology,
21  (2) how the biology is perturbed by toxicants,
22  and (3) how and when adaptive and adverse responses develop. Sufficiently well-developed and
23  well-validated models can be used to predict dose-response and time course behaviors for the
24  perturbations, adaptive responses, and apical health effects, but the accuracy of these predictions
25  depends on the extent and quality of the data used as inputs and on the technical quality of the
26  model itself. Time-course and dose-response pathway data from *in vitro* exposure studies can be
27  paired with pathway data from *in vivo* exposure studies and assembled into a nodes-and-edges
28  graph encompassing mechanisms of action relevant to ozone toxicity, focusing on pathways most
29  relevant to lung inflammation. This pairing and assembly will provide a framework for modeling
30  ozone toxicity pathways to downstream pathophysiological changes (see Figure 9, event 3). At the
31  intracellular level, upstream signaling pathways (e.g., NF-κB) that have been shown to mediate
32  ozone-induced changes in gene expression will be represented, connecting the oxidative products
33  of ozone formed in the cell to time-dependent changes in protein activity and RNA expression. For
34  example, the canonical NF-κB signaling pathway shown in Figure 9 plays a role in ozone-induced
35  inflammation. Finally, data on ROS production resulting from ozone exposure (see Figure 9, event
36  1) will be represented in the model, both as an input to ozone's perturbation of the molecular-level
37  components and as drivers of downstream signaling pathways.

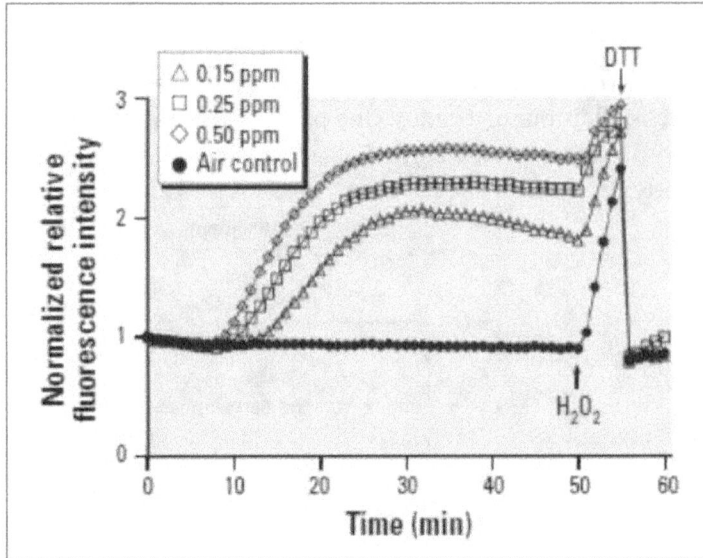

Figure 13. Exposure to ozone induces a rapid increase in intracellular reactive oxygen species (ROS). Addition of 0.1 mM $H_2O_2$ at the end of the ozone exposure produced a maximal response, which was fully reversible with the addition of 10 mM dithiothreitol (DTT), a strong reducing agent (Gibbs-Flournoy et al. 2013). Reproduced with permission from *Environmental Health Perspectives*.

## Susceptibility

15  Not all individuals are equally
16  responsive to toxicants; some are
17  much more responsive because of
18  age, gender, disease, lifestyle (e.g.,
    obesity), or genetic/epigenetic
19  factors. For example, the range of
    response in lung function
    decrements to ozone in young
    healthy individuals (McDonnell et al.
    2012) is 10-fold. Individuals exposed
    to ozone a second time, many
    months later, retain their hierarchy
    on the response curve, implying that
    a long lasting factor, perhaps genetic
    or epigenetic, plays a role in ozone
    responsiveness. Asthmatics are
    known to have an enhanced
    inflammatory response to ozone
    (Bosson et al. 2003, Peden et al.
20  1997), as do individuals carrying the
21  GSTM1 null allele (Kim et al. 2011).
22  Understanding the MOA by which a

23  person is more responsive to a pollutant should be a component of a systems biology approach to
24  toxicity testing. Airway epithelial cells can be obtained from more-responsive and less-responsive
25  individuals, and the pathways altered by ozone can be compared for both groups. Recently,
26  cultured lung epithelial cells obtained from individuals carrying the GSTM1 null allele have been
27  shown to be more responsive to air pollutants than cells obtained from individuals carrying the
28  wild-type GSTM1 allele (Wu et al. 2011). Airway epithelial cells obtained from asthmatics appear to
29  retain an asthma phenotype in culture and are more responsive to pollutants than cells obtained
30  from nonasthmatics (Duncan et al. 2012). Readily obtaining bronchial airway cells can be difficult,
31  so knowing that the response of cultured nasal epithelial cells to toxicants has recently been shown
32  to be similar to that of bronchial cells (McDougall et al. 2008) can be instructive. These nasal cells
33  can be readily and noninvasively obtained from most individuals, including children.

## Involvement of the Inflammatory Network in Multiple Diseases

34  Chronic inflammation is implicated in the etiology of several diseases, including atherosclerosis,
35  heart disease, obesity, diabetes, arthritis, cancer, and lung diseases (asthma, emphysema,
36  pulmonary fibrosis). Both common and disease-specific inflammatory molecular patterns have
37  been reported to underlie these diseases (Wang, I et al. 2012). Why a particular disease is
38  expressed in an individual or subpopulation as the result of inflammation is likely the result of the
39  site of injury, co-activation of other networks, genetic variation, or environmental factors. Such
40  complicating factors therefore highlight several issues that might arise when using molecular
41  patterns to predict disease risks: (1) observation of an inflammatory disease signature for a
42  chemical that has not been well studied would raise concerns for inflammatory disease risks; (2)
43  the specific inflammatory disease in question likely would be difficult to predict with a limited

1 systems biology context; (3) a network might be involved in multiple diseases; and (4) the specific
2 disease expressed could involve multiple interactive pathways and networks.

3 Specifically, many air pollutants appear to induce cardiopulmonary inflammation, which likely
4 plays a role in risks for asthma, emphysema, and pulmonary fibrosis. Molecular biology is likely to
5 be a useful tool in sorting out the relative contributions of various air pollutant exposures to
6 cardiopulmonary disease via inflammatory mechanisms.

## Risk Assessment Implications Based on the Ozone Prototype

7 *Hazard Identification* – The pathway information, coupled with data about ozone-induced changes
8 in upstream transcription factors, signaling pathways, and generation of ROS, can lead to the
9 development of molecularly based dose-response system models that are predictive of downstream
10 *in vivo* pathophysiological changes. These data suggest that ozone activates the NF-κB and ERK
11 pathways, both known to modulate inflammation, *in vitro* and *in vivo*. This suggests that the *in vitro*
12 airway epithelial cell model used here might be amenable to predicting *in vivo* inflammation. An
13 HTS assay based on this cell model might be able to provide rapid hazard identification in the
14 future.

15 *Exposure-Dose-Response Assessment* – We did not perform an analysis of transcriptional changes
16 across a range of doses.

17 *Cumulative Risk Assessment* – This *in vitro* model could be used to make comparisons of the
18 transcriptional response upstream of the inflammation process using complex mixtures of air
19 pollutants. The comparison, however, might require specialized equipment and monitoring to
20 ensure the mixture and dose of pollutants are proper and well controlled.

21 *Variability and Susceptibility in Human Response* – In the future, this and other similar models
22 might identify pathways and mechanisms by which susceptible human populations respond to
23 inhaled toxicants. Just as this *in vitro* model was derived from several young, healthy volunteers,
24 performing a larger study of variability and susceptibility would be possible by recruiting and
25 including specific populations. Such a study also would facilitate the creation of HTS assays for
26 rapidly studying susceptible populations and variability in response.

### 3.1.3. Benzo[a]pyrene (a Polycyclic Aromatic Hydrocarbon), and Cancer

27 PAHs are produced from combustion or pyrolysis of carbon-containing material, exist in the
28 environment almost exclusively as complex mixtures, are a major component of urban air pollution,
29 and are a drinking water contaminant. Several PAH-containing complex mixtures are known to be
30 carcinogenic in humans (e.g., coke oven emissions, diesel exhaust, and tobacco smoke). Many
31 individual PAHs and PAH-containing mixtures have been tested in traditional bioassays; many, but
32 not all, appear carcinogenic. Additionally, those that are carcinogenic vary in terms of potency.
33 Given the universe of PAHs and potential PAH-containing mixtures, testing them all is not feasible.
34 Hence, an alternative approach using molecular biology was explored in this prototype. See Text
35 Box 7 for some challenges related to this prototype.

36 This effort focused on one PAH—B[a]P—and liver cancer. Repeated B[a]P exposure has been
37 associated with increased incidences of total tumors and of tumors at the site of exposure (dietary,
38 gavage, inhalation, intratracheal instillation, and dermal and subcutaneous, in studies of numerous

strains and species of rodents and several nonhuman primates). Distant site tumors also have been observed after B[a]P administration by various routes, and B[a]P is frequently used as a positive control in carcinogenicity bioassays.

## Systems Biology Model

EPA (2013), Burgoon (2011), have proposed a cellular systems model and pathways based on a systematic meta-analysis of transcriptomics data for B[a]P-mediated liver cancer (Figure 14 and Table 2). The core of the model is focused on induction of DNA adducts, mediation of p53 (a tumor suppressor gene) signaling, alterations of translesion synthesis,[11] and regulation of the G1/S-phase transition and the cell cycle. Based on this model, the DNA adducts are believed to be formed by reactive B[a]P metabolites through cytochrome P450 (CYP) enzyme induction, secondary to B[a]P activation of the aryl hydrocarbon receptor (AhR). Others have shown AhR-independent DNA adduct formation, raising questions about other non-CYP1A1- and CYP1A2-mediated B[a]P metabolism and adduct formation (Sagredo et al. 2006, Kondraganti et al. 2003).

The systematic meta-analysis started with a search for published, peer-reviewed transcriptomics data sets using B[a]P as the test substance. The Gene Expression Omnibus (GEO) and ArrayExpress databases were searched for microarray transcriptomic studies using the search terms in Table 3. The search focused on GEO and ArrayExpress as these databases store submitted data as raw data. The raw data are critical for performing meta-analyses, especially when different analysis methods might be used.

> **Box 7. Challenges Encountered With This Prototype**
>
> This prototype originally focused on identifying whether human transcriptomics data from PAH mixtures found in cigarette smoke could be associated with lung cancer. This prototype was envisioned as a real-world example of how data mining of existing data could be informatively performed. Unlike the other Tier 3 Prototypes, which were designed to have the best combination of data available, however, this prototype encountered numerous data access and experimental design challenges that we expect to be seen when applying these methods in the future. These challenges included:
>
> - An inability to easily obtain the raw data required for re-analysis of the transcriptomics data.
> - Lack of clear descriptions of the study design or analysis method.
> - Different microarray platforms being used.
> - Different analysis methods being employed within the same platform.
> - Lack of a quantitative exposure estimate (especially common with human studies that lack a controlled exposure).
>
> Together, these challenges make performing a quantitative meta-analysis difficult. For new types of data to be useful, improvements to data collection and concomitant exposure analyses are needed.

---

[11]Translesion synthesis is a mechanism that the cell uses to continue DNA replication/synthesis in the presence of a DNA lesion (e.g., DNA adduct).

*This document is a draft for review purposes only and does not constitute Agency policy. Do not cite or quote.*

September 2013                                          34

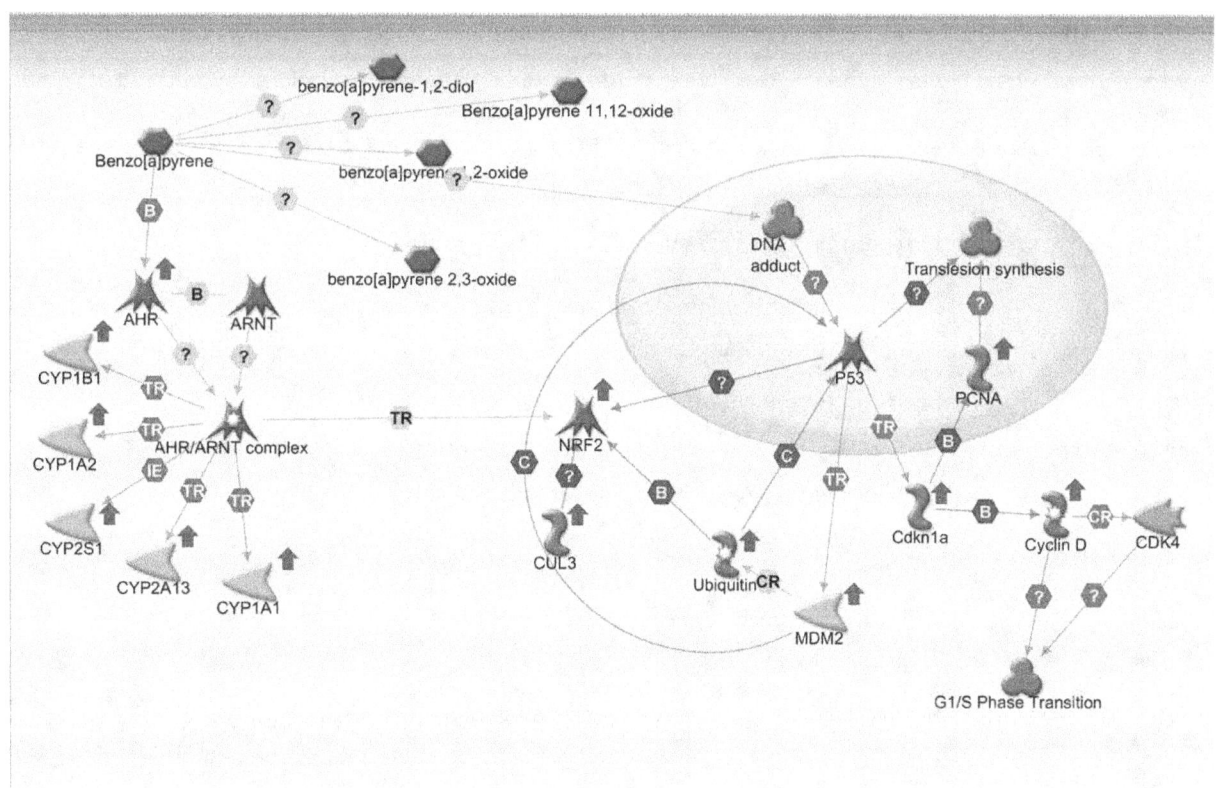

Figure 14. Consensus Outcome Pathway. This consensus pathway was synthesized by combining multiple pathway diagrams identified through analysis of the two data sets using GeneGo Metacore. The nodes (proteins or outcomes) are connected by lines. The green lines represent activation, while the red lines represent inhibition or repression. The thick red arrows near proteins represent increases in gene expression.

1  The search resulted in the identification of 26 peer-reviewed publications with 40 gene expression
2  data sets. The adult mouse liver was chosen as the focus system based on the number of studies
3  available across the species and tissues where B[a]P was used. Only 2 of the 26 publications
4  focused on *in vivo* transcriptomic studies of the liver in the mouse. Study GSE24907 is a dose-
5  response study where five male Muta mice (a LacZ transgenic mouse line) per group were gavaged
6  with an olive oil vehicle and 25, 50, or 75 mg/kg B[a]P. Study GSE18789 is a time-course study
7  where 27- to 30-day-old B6C3F1 mice were gavaged with 150 mg/kg B[a]P for 3 days and
8  sacrificed at 4 or 24 hours after the final dose.

## Table 2. Altered Genes/Functions and Their Relationship to Cancer (in this Model)

| Altered Gene or Function | Relationship to Cancer in this Model |
|---|---|
| AhR/ARNT Complex | AhR regulated expression of several CYPs, including CYP1A1 and CYP1A2 |
| CYPs (e.g., CYP1A1, CYP1A2) | Upregulation leads to production of oxidative radicals and B[a]P metabolites |
| NRF2 | Regulates the expression of oxidative stress-protective genes |
| Ubiquitin | Protein that tags other proteins for destruction |
| CUL3 | Regulates the inhibition of NRF2 signaling with ubiquitin |
| p53 | Stops cell cycle by preventing G1/S phase transition; activated by DNA damage |
| MDM2 | Regulates p53 through negative feedback mechanism with ubiquitin |
| Cdkn1a/p21 | Upregulated by p53 activation; inhibits Cyclin D activation and prevents G1/S phase transition |
| Cyclin D | Activates G1/S phase transition, works with CDK4 |
| CDK4 | Activates G1/S phase transition, works with Cyclin D |
| G1/S Phase Transition | Starts cell cycle progression by allowing for DNA synthesis |
| Translesion Synthesis | DNA damage tolerance mechanism; allows DNA replication fork to progress beyond DNA damage sites |
| DNA Adduct | A piece of DNA covalently bound to a chemical that can modify expression of DNA |

## Table 3. Search Terms and the Number of Studies Retrieved from the Gene Expression Omnibus (GEO) and Array Express Microarray Repositories

| Search Term | Number of Microarray Studies Retrieved |
|---|---|
| Coal tar | 2 |
| Polycyclic aromatic hydrocarbons or PAHs | 13 |
| Diesel | 11 |
| Smoke (NOT cigarette smoke) | 16 |
| Benzo[a]pyrene or B[a]P | 53 |
| Fuel oil | 1 |
| Cigarette smoke | 63 |
| Tobacco smoke | 16 |

1 The Systematic Omics Analysis Review (SOAR) Tool was used to document and facilitate the evaluation of both studies (McConnell and Bell 2013). SOAR consists of 35 objective questions that help users determine if a study contains data of sufficient quality for use in a risk assessment context. SOAR was developed by toxicology and toxicogenomics experts, and based, in large part, on existing and published data standards such as the Minimum Information About a Microarray Experiment (MIAME) standard. Both studies (GSE24907 and GSE18789) met the SOAR screening threshold. Following a more in-depth scientific review, both studies were found to be of sufficient quality for use.

1    That DEG lists reported in the peer-reviewed literature are not reproducible across similar studies
2    is well established (Shi et al. 2008, Chuang et al. 2007, Ein-Dor et al. 2005, Lossos et al. 2004,
3    Fortunel et al. 2003). In one published example, three different studies aimed at identifying
4    "stemness" genes[12] each yielded 230, 283, and 385 active genes, yet the overlap between them was
5    only 1 gene (Fortunel et al. 2003). Therefore, a pathway-based meta-analysis approach was used,
6    whereby fold change-based ranking, or more formal meta-analyses relying on raw data, along with
7    a standardized analysis approach are considered to be more reproducible than published DEGs
8    (Ramasamy et al. 2008, Shi et al. 2008, Chuang et al. 2007).

9    Both studies were reanalyzed independently at the feature level[13] using the same pre-processing,
10   normalization, and analysis methods. GeneGo Metacore was used to identify pathways representing
11   a large number of genes from both data sets.

12   The consensus systems model (Figure 14) was synthesized based on the results from GeneGo
13   Metacore. The model conceptually describes the events that might occur when B[a]P enters the cell.
14   Briefly, B[a]P binds to AhR, leading to upregulation of xenobiotic metabolizing enzymes and Nrf2,
15   which might lead to additional B[a]P metabolism to epoxides and increased oxidative stress.
16   B[a]P-mediated genotoxicity, evidenced by DNA adducts, will occur and will activate p53. Although
17   Nrf2 is upregulated transcriptionally, p53 is expected to interfere with Nrf2 signaling, ensuring a
18   pro-oxidant environment, which might perpetuate further DNA adduct formation. Upregulation of
19   p21 (Cdkn1a) and MDM2 are most likely a result of p53. Upregulation of ubiquitin, while in the
20   presence of p53-mediated MDM2 upregulation, is expected to destabilize p53. Destabilization of
21   p53, in the presence of PCNA, is expected to allow translesion synthesis, which will allow mutations
22   and adducts to perpetuate through DNA synthesis. Upregulation of Cyclin D could be sufficient to
23   overcome p21 inhibitory competition, especially as p53 levels decrease, allowing for G1/S phase
24   transition to occur. Thus, G1/S phase transition, combined with translesion synthesis, is expected to
25   lead to propagation of mutations and DNA adducts into daughter cells. This loop might continue
26   into a feed-forward situation until p53 signaling can be reinitiated.

---

[12]"Stemness" genes are those genes that are hypothesized to confer stem cell characteristics.

[13]A common misconception about microarrays is that they measure gene expression at the level of a gene. In reality, microarrays measure only a portion of a gene, typically anywhere from 20 to 100 nucleotide bases. This portion of the gene that is actually measured is called a "feature." Typically, only one feature exists per gene on a microarray. Some genes are represented more than once on a microarray, however, complicating downstream analyses (e.g., deciding how much a gene is expressed when the two features representing different parts of the same gene yield different numbers). Features could also be believed to map to a specific gene at one time, and the feature is later discovered to map to a completely different gene (this happens more frequently with lesser known or studied genes and lesser known or studied organisms where the genome might not be available). Thus, the gene associated with a feature can change over time, and most analysts will re-map their feature sequences against the genome periodically to ensure they have the latest annotation. This might result in reproducibility issues when comparing to studies performed at different times. Generally, when interpreting gene expression, analysts prefer to operate at the feature level for all analyses.

*This document is a draft for review purposes only and does not constitute Agency policy. Do not cite or quote.*

September 2013                 37

1  Using the gene expression changes and activating DNA adduct formation, the Boolean Network
2  systems model (Figure 15-17)[14] predicts that cell cycle progression will be activated with
3  translesion synthesis[15] (Figure 18). These data and the systems model support the notion that the
4  high doses and acute durations used in the two mouse liver studies might initiate liver tumor
5  progression through a genotoxic MOA, and promotion might occur through a cellular proliferation
6  MOA. Due to the lack of data, speculating whether this system could be activated at low doses in the
7  mouse is not possible. Due to genetic and epigenetic variability and potential species differences,
8  these types of effects might occur at lower doses in humans than in mice.

9  The proposed model, however, provides a testable hypothesis for effects at lower doses, with other
10  species, and other PAHs. For instance, transcriptomic studies with PAH mixtures, or other PAHs
11  individually, can be analyzed to see if they might also impinge on this pathway. Further, the gene
12  expression data from these other studies can be placed into this model, and an analysis can be
13  performed to see how the cell might react, compared to B[a]P. This will give an indication of
14  doses/exposures that could lead to DNA damage, activation of translesion synthesis, and G1/S-
15  phase transition.

## Human Susceptibility and Population Variability

16  Variations in human genetics will alter the susceptibility and population variability with respect to
17  the tumorigenesis or carcinogenesis outcomes. For instance, SNPs are known to occur in p53, which
18  might impact its ability to stop G1/S phase transition. In addition, the p53 gene has been shown to
19  be mutated in many cancers (Vogelstein et al. 2000). A data mining approach can be taken to
20  identify other relevant SNPs for the genes or proteins in the systems model.

---

[14]In a Boolean Network model, the system is represented as a series of connected nodes. Each node represents a gene/protein, and a connection represents some type of action/inhibition relationship. The connections are directed. For instance, p21 inhibits Cdk4, so the arrow originates at p21 and terminates at Cdk4. Some of the relationships are not as direct. For instance, Cyclin D interacts with Cdk4 to activate G1/S phase transition; however, in the model, this is best represented as a positive interaction between Cyclin D and Cdk4 given the relationship between Cdk4, Cyclin D, and p21. Each node has a state, either on (1) or off (0). Based on the state and the relationship to the other nodes, the Boolean Network can cycle through a series of states. To test the predicted outcomes (i.e., can the model sustain cell cycle progression and translesion synthesis once initiated?), this model was further simplified into just the DNA adduct/cellular proliferation part, and represented as a Boolean Network systems model. Specifically, we are looking for stable states or attractors—cycles of states that recur and self-perpetuate. States that lead to attractors are called the basin. The Boolean Network in Figure 15 has a single state attractor defined in Figure 16. This state can be defined as a cell cycle progression state with translesion synthesis turned on. If the cell were to enter this system state, it would be expected to self-perpetuate until a stimulus shuts it down. Important to note is that the systems model does not predict that all cells will enter this state or that this state is the default. Rather, the model is simply stating that if this state were entered, the cell would remain in this state until a stimulus occurs that forces it out. Such stimuli might include changes in gene expression, alterations of metabolic states, or a change in overall energy level. The Boolean Network model predicts that, with DNA adducts alone, the cell will enter into a five-state attractor (Figure 17). In this cycle, the cell is not predicted to enter into G1/S phase transition—which is expected because p53 should effectively shut down that pathway. Translesion synthesis is predicted to occur in this attractor cycle.

[15]Translesion synthesis is a mechanism for DNA damage tolerance that allows the DNA replication machinery to move beyond a DNA lesion or abasic site (i.e., a site that lacks a DNA base).

---

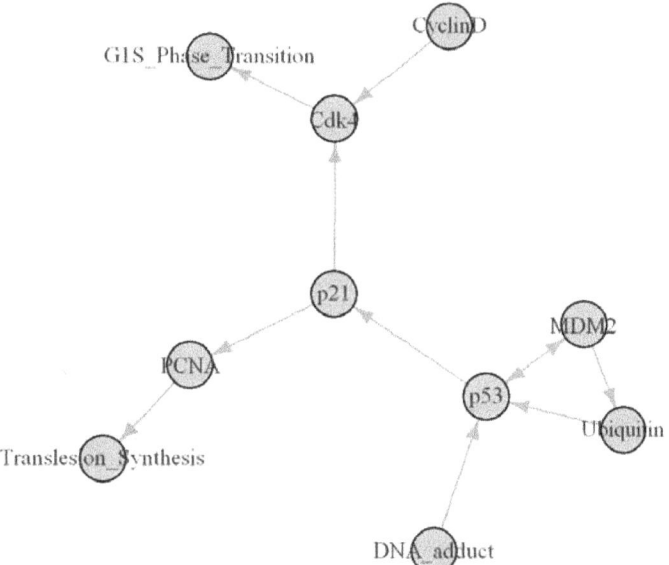

Figure 15. Liver Carcinogenesis Systems Model. The nodes represent proteins, and the lines are directional connections meaning activation or inhibition (activation and inhibition are not treated differently in the graphical depiction of the model). For instance, the arrow from PCNA to translesion synthesis means that PCNA activates translesion synthesis. The two major outcomes in this model are translesion synthesis and G1/S phase transition. The major external input is DNA adduct formation. DNA adducts cause structural damage to the DNA, which could become or lead to mutations and ultimately tumorigenesis and cancer.

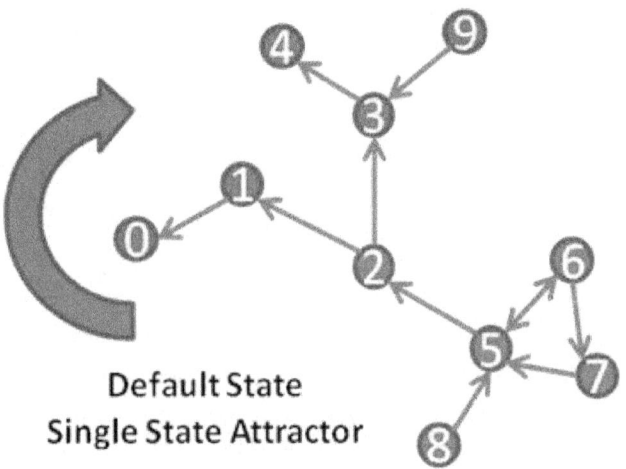

**Default State**
**Single State Attractor**

| | |
|---|---|
| 0: Translesion Synthesis |
| 1: PCNA |
| 2: p21 |
| 3: Cdk4 |
| 4: G1/S Phase Transition |
| 5: p53 |
| 6: MDM2 |
| 7: Ubiquitin |
| 8: DNA adducts |
| 9: Cyclin D |

Figure 16. Default State, Single State Attractor. The systems model falls into a default state, single state attractor system. This is the same as the network represented in Figure 15. The names have been replaced by numbers, which are noted in the figure legend. Red nodes are those that are activated. Blue nodes are inactivated. The system here has not been perturbed by external forces. Of particular interest is that the "default" state for the system is one where the cell is actively proliferating and undergoing translesion synthesis.

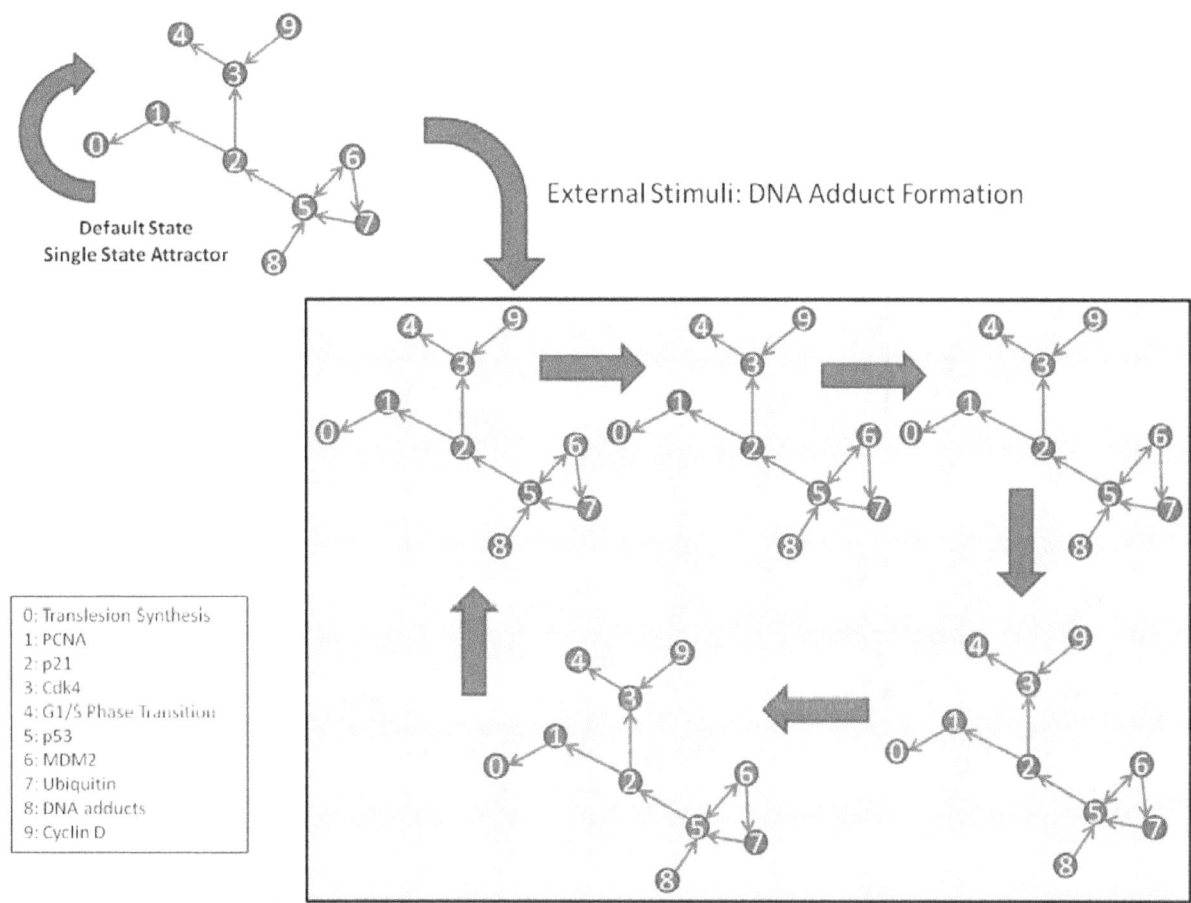

Figure 17. DNA Adduct Attractor System. When the systems model is perturbed through an external stimulus (DNA adduct formation), it transitions from the default stable starting state and moves to a new attractor (depicted in the inset). Once the system moves out of the basin for the default state attractor, it cannot return to that state without another significant stimulus. This multistability (the fact that a system can have multiple stable attractor states) is a characteristic of complex systems. Starting at the upper left of the inset, PCNA is activated, DNA adducts are activated, and p53 is activated. This leads to translesion synthesis and activation of p21, MDM2, and ubiquitin. Although Cyclin D gets activated, there is no activation of G1/S phase transition. The system then transitions to a state where translesion synthesis is primed and ready to go. If G1/S phase transition were to occur, p53 is activated, along with DNA adduct formation, MDM2, and ubiquitin. The next system state has continued p21 activation, loss of p53 activity presumably through ubiquitin and MDM2 activation in the prior system state, and DNA adduct formation. The system then transitions to only DNA adduct formation and ubiquitin activation, followed by restarting of the cycle.

*This document is a draft for review purposes only and does not constitute Agency policy. Do not cite or quote.*

September 2013           41

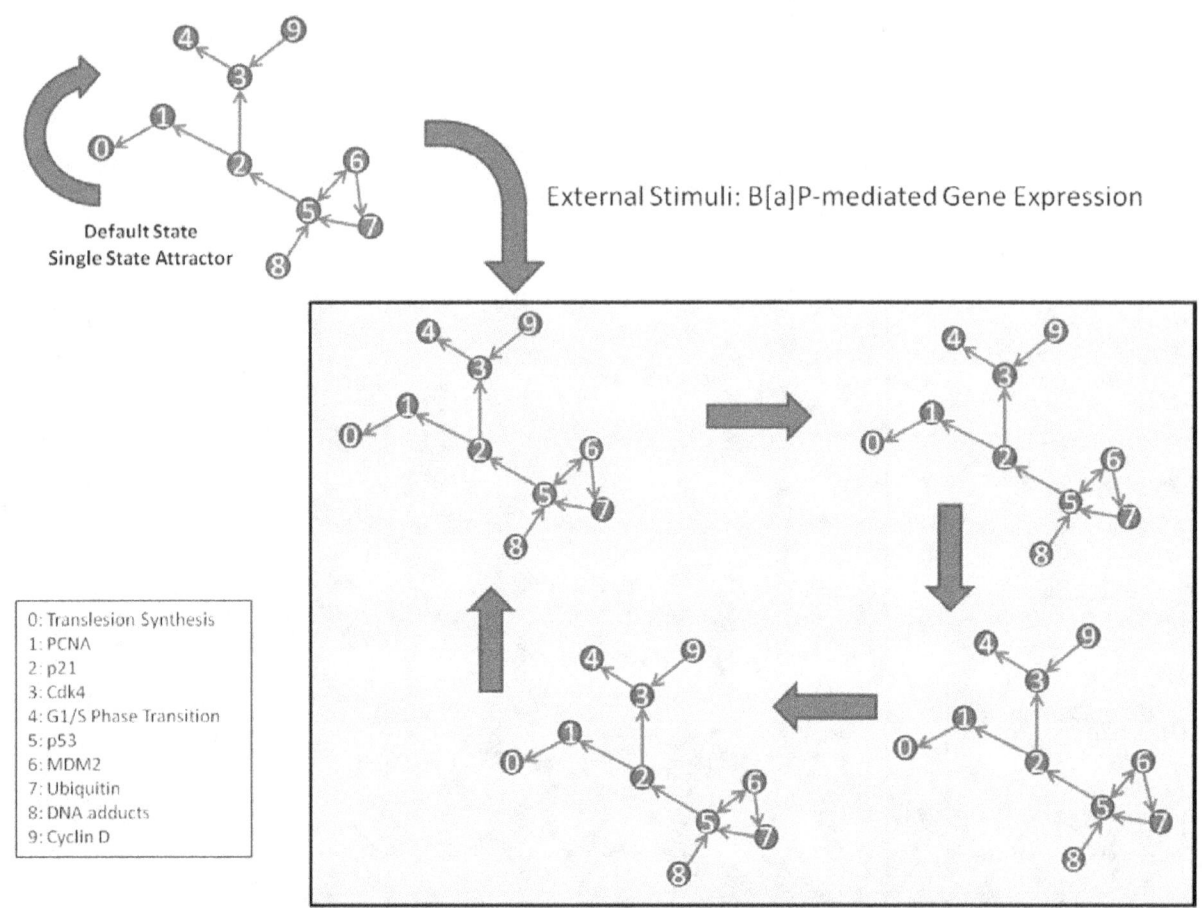

External Stimuli: B[a]P-mediated Gene Expression

Default State
Single State Attractor

0: Translesion Synthesis
1: PCNA
2: p21
3: Cdk4
4: G1/S Phase Transition
5: p53
6: MDM2
7: Ubiquitin
8: DNA adducts
9: Cyclin D

Figure 18. Gene Expression Data Attractor System. This four-system attractor is based on the gene expression data observed in both studies. This attractor system is notable as it shows DNA adduct formation, translesion synthesis, and G1/S phase transition occurring in all system states. This model predicts that DNA adducts and potential mutations are being passed forward to daughter cells through translesion synthesis as the cell cycle progresses at these doses and times in the mouse liver. This suggests that B[a]P at these doses and experimental time-points post exposure in the mouse liver could be an initiator and promoter of tumorigenesis. This adverse outcome pathway (AOP) might ultimately result in carcinogenesis.

## Risk Assessment Implications Based on the B[a]P Prototype

1  *Hazard Identification* – These data suggest that B[a]P activates known human disease pathways
2  associated with genotoxicity and tumor promotion/cell cycle progression. Similar pathway-based
3  meta-analyses can be performed on transcriptomic data for other chemicals to screen for
4  genotoxicity and tumor promotion, prior to the observation of tumors. For instance, using this
5  specific Boolean systems model would inform risk assessors of the likelihood that other PAHs and
6  PAH mixtures share a similar AOP. This type of chemical screening would need to be further
7  validated with known or likely carcinogens and compared against chemicals that are believed not
8  to be carcinogens (to establish performance of the screening method).

9  Disease-focused system models could be developed for a larger set of complex human diseases to
10  expand the utility of this approach in the future. The pathway-based, diseased-focused, Boolean

1 systems model approach could be expanded to include emerging data streams, including
2 metabolomics and proteomics, to create overall improvements in mechanistic understanding and
3 hazard identification screens.

4 The genes in these Boolean systems models can be considered as those that might be tested
5 together in a battery of assays to be used in Tox21 screening. HTS assay batteries based on these
6 models can be implemented easily using current multiplex quantitative PCR assay systems.

7 *Exposure-Dose-Response Assessment* – Analyzing changes in the systems model and potential
8 differences in adverse outcome across a range of doses was precluded due to the lack of sufficient
9 dose-response data. The Boolean systems models approach used here, however, would allow for
10 the prediction of adverse outcomes across a range of doses. In the B[a]P example, we examined the
11 impacts of different scenarios. This same approach would be used to analyze different doses.

12 *Cumulative Risk Assessment* – Boolean systems models can be used to compare and integrate
13 pathway-based results from multiple chemicals and nonchemical stressors. This approach would
14 enable prediction of hazards from exposure to mixtures or cumulative stressors.

15 *Variability and Susceptibility in Human Response* – Human susceptibility can be modeled by
16 using data from genome-wide association studies (GWAS), knock-out studies, or knock-down
17 studies. In this instance, modeling the impacts on the adverse outcome predicted by the Boolean
18 systems model is possible. For instance, the impacts of a gene knock-out generally can be modeled
19 in the Boolean systems model as a constant inactivation of the protein.

20 Population variability would be modeled using a Monte Carlo simulation to estimate the risk of
21 adverse outcomes across different genetic profiles. This would be accomplished by using the same
22 types of models as in the human susceptibility context. The population variability scenario can be
23 considered as creating a population of susceptibility Boolean systems models, where each model
24 has a chance of being included in the overall analysis equal to its occurrence in the human
25 population (or equal to its occurrence in a hypothesized human population if performing a what-if
26 type of scenario). For instance, if 15% of the population is expected to have a loss of function
27 polymorphism, the Monte Carlo model should have a 15% chance of choosing that type of Boolean
28 systems model on each random draw from the population.

### 3.1.4. Risk Assessment Implications across the Tier 3 Prototypes

29 Looking across the Tier 3 prototypes:

30 • Benzene, ozone, and B[a]P displayed human molecular signatures that are strongly associated
31 with specific human disorders and diseases.

32 • This type of molecular mechanistic understanding can be used to screen and predict an
33 association between a chemical and a disease, or to augment the existing weight of evidence
34 for an association between a chemical and a disease.

35 • With sufficient systems biology understanding and data, disease signatures also could be used
36 to screen chemicals with no or limited traditional data for specific disease hazards.

- Meta-analyses that integrate pathway-based data across multiple studies yield the greatest evidence that associate chemical exposure to a disease, and are generally the most appropriate method for using transcriptomics data in a risk assessment. A pathway analysis from a single study will yield more evidence to associate a chemical exposure to a disease, and assuming the study design is adequate, might be appropriate for a risk assessment. An analysis built on a set of DEG lists is not reproducible or adequate for risk assessment purposes.

- Dynamic disease-based systems models will facilitate the understanding and prediction of chemical-disease associations in the near future. These models provide a nonbiased view of the underlying biology, and can facilitate making pathway-based predictions of adverse outcomes and disease when the interconnections within the pathway become complicated (e.g., the B[a]P case study).

- On an individual level, molecular signatures involve dynamic relationships among adaptive and nonadaptive processes that will require additional research to understand fully. At the population level, environmental factors can be thought of as shifting the population or subpopulation distributions toward (e.g., certain chemical exposures) or away from increased levels of risk (e.g., beneficial nutrients).

- *In vitro* responses appear to have commonalities with *in vivo* responses but also are affected by a number of variables, such as test system, metabolism, cell type, tissue type, time course of events (ozone data only), individual characteristics (intrinsic and extrinsic), and species.[16] These complexities make the identification of a specific disease hazard from *in vitro* only data difficult. Systems biology understanding, derived from *in vivo* data, increases confidence in the interpretation of *in vitro* data.

- For *in vitro* data, identifying hazards that occur at the organ or organismal level might be difficult. Thus, *in vitro* studies might be more appropriate for assessing the relative potencies of chemicals to alter biological processes (vs. induce disease) or to predict hazards that occur or are initiated at the tissue level (e.g., generalized inflammatory response). This is particularly true if relative potency is evaluated within a given protocol.

- Future research merging GWAS data and personalized medicine into organized data can help better characterize both intrinsic and extrinsic factors that contribute to human variability and susceptibility.

- The networks associated with a disease can apparently be disrupted in multiple places, all leading to altered risks of the specific disease. This is shown by mechanistic commonalities among diseases of unknown origins, other chemicals associated with the disease, and chemotherapeutics that can reverse or block components of the disease processes. This type of information can be a useful tool in characterizing cumulative risks. Overly narrow descriptions of mechanisms can miss interactions among environmental factors.

---

[16]Although not evaluated here, lifestage is also an important variable (Boekelheide et al. 2012).

1   &bull;    When searching for candidate Tier 3 prototypes, one important observation was that, even
2        among the most well studied chemicals, very few chemicals had the type and quality of data
3        needed for exploring the use of new data types in risk assessment. There are needs for
4        systematic review criteria for new data types, adherence to standards of experimental and
5        statistical practices in data generation and analyses, and thoughtful consideration of
6        variability and uncertainty to improve the utility of new data types for risk assessment.

## 3.2. Tier 2: Limited Scope Assessments

7 The intent of the Tier 2 prototypes is to (1) explore new types of computational analyses and
8 short-duration *in vivo* bioassay that are currently relatively uncommon in risk assessment but hold
9 great promise for the near future; and (2) develop an assessment approach well suited to limited
10 scope risk management decisions. In this case, "limited" generally means regional to local exposure
11 potential, or limited hazard potential, or limited data to conduct more detailed assessments. Tier 2
12 efforts fall between Tier 3 and Tier 1 in terms of resources required and amount of uncertainty in
13 the assessment results. The number of chemicals possibly identified in Tier 1 meriting further
14 testing could overwhelm traditional or Tier 3 type evaluation, thus the need for an intermediate
15 testing and assessment strategy as provided by Tier 2 (Thomas, RS et al. 2013a).

16 The hallmark of Tier 2 data in the NexGen program is integration across biological systems—
17 molecule-to-cell(s)-to-tissue(s) and, in some systems, to-outcome(s)—to inform associations
18 among environmental exposures, causal mechanisms, and outcomes, but generally using
19 evaluations over relatively short time periods (hours to weeks). Tier 2 considers all information
20 available from Tier 1 approaches, such as quantitative structure-activity relationship (QSAR) and
21 HTS, along with other data derived from more complicated test systems that use intact tissues or
22 organisms to provide a higher level of confidence in the assessment (Table 4). Limited scope
23 assessment could include combining HTS with limited traditional data. Tier 2 data are commonly
24 referred to as high-content data.

**Table 4. Summary of Tier 2 NexGen Approaches, Including Weight of Evidence, Pros, and Cons**

| | Tier 2: Limited Scope Assessments Categories of Approaches Considered | | |
| --- | --- | --- | --- |
| | Data Mining of Existing Databases | Alternative Species *In Vivo* Assays | Mammalian Short-duration *In Vivo* Assays |
| **Approach:** | Discovers or identifies associations among environmental exposures, omic patterns, and human disease. Often uses meta-analyses of large existing data sets. Suggests potential adverse outcomes based on existing knowledge of other chemical-induced molecular event and disease relationships. | Experimentally measure dose-dependent, chemically-induced alterations in biological functions in intact organisms using a range of specific and sensitive assays. Measures adverse outcomes that range from omics to phenotypic outcomes and population effects. | Experimentally measure dose-dependent, chemically-induced alterations in biological functions in intact animals using a range of specific and sensitive assays. Measures molecular or cellular changes; infers potential adverse outcomes based on existing knowledge of other chemical pathway or disease relationships. |
| **Weight of evidence:** | Determined by the quality and amount of underlying evidence (ranges from suggestive to likely) or is known with substantial complementary experimental data. | Determined by the quality and quantity of data, but generally suggestive to likely. Cross-species issues need consideration. | Determined by the quality and amount of underlying evidence, ranges from suggestive to likely when anchored to pathway and traditional data and some understanding of temporal progression. |
| **Pros:** | Significantly faster and less expensive than traditional bioassays. Can use combined data sets that include tens of thousands of humans. Includes tissue, organism, and life span-level integration, including metabolism | Significantly faster and less expensive than traditional bioassays. Can evaluate complex outcome birth defects and neurobehavioral outcomes. | Significantly faster and less expensive than traditional bioassays. Includes tissue and organism integration, including metabolism. |
| **Cons:** | Relationships generally associative; might be causal in certain circumstances (depending on data quality and amount of underlying evidence). Data on effects of early life exposures and effects generally lacking. | Species-to-species extrapolation is an issue as is the potential for parent compound not to be metabolized to toxicants that are active in humans. Omics information can be derived from organs, tissues, and multiple cell types versus only human-based target cells. Data on effects of early life exposures and effects generally lacking; an exception is the embryonic fish models. | Measure events early in disease initiation process; early events could be reversible; links to apical outcome can be an issue. Omic information is often derived from multiple cell types versus only target cells. Data on effects of early life exposures and effects generally lacking. |

1  Two general approaches to Tier 2 data are discussed here:

2  • High-content knowledge mining (i.e., computer-driven surveys of the literature and large
3     existing data libraries[17]) to retrieve data and conduct meta-analyses of existing systems
4     biology data to construct mechanism-of-action models and establish associations between
5     environmental exposure and disease. The diabetes/obesity prototype is provided as an
6     example.

7  • Short-term *in vivo* or *in situ* exposures of intact organisms to enable incorporation of the
8     intact metabolism in the toxicity evaluation and produce measures of biological change over a
9     short time frame (i.e., ranging from hours to a few months) that are thought to be relevant to
10    longer term outcomes. Two examples are provided using alternative and mammalian species.
11    Considerable work is ongoing at various U.S. Federal Government agencies and elsewhere to
12    refine assays where animals are exposed to chemicals *in vivo* for periods ranging from hours
13    to a few months.

14  Implications for risk assessment identified by the Tier 2 prototypes are discussed at the end of this
15  section and integrated with other lessons learned in Section 5 "Lessons Learned from Developing
16  the Prototypes."

### 3.2.1. Knowledge Mining – Diabetes/Obesity

17  Knowledge mining[18] is explored in this prototype as a means to characterize the associative and
18  potentially causal relationships among disease and exposures to environmental factors and
19  intrinsic sources of human variability. The knowledge mining approach capitalizes on huge new
20  databases that are being supplemented with each publication in the field of omics (>50,000 per
21  year). These databases are generally oriented toward the omics of human disease but also include
22  omics information on other species, as well as surveys and clinical assays measuring human
23  exposure and health outcomes. The specific, related diseases explored here are diabetes and
24  obesity and relationships to multiple environmental factors. Diabetes results from environmental
25  and genetic factors and risk varies considerably in the population (Patel et al. 2013). Four
26  interrelated efforts focusing on diabetes/obesity are reported here: (1) Comparison of Knowledge
27  Mining Results and Expert Opinion; (2) Environment-wide Association Studies (EWAS); (3) Itemset
28  Associations between Prediabetes/Diabetes and Chemical Exposures; and (4) Characterizing
29  Human Susceptibility and Population Variability.

### Comparison of Knowledge Mining Results and Expert Opinion

30  Thayer et al. (2012) reported on a recent National Toxicology Program (NTP) workshop that
31  examined the possible causal relationships between environmental exposures and diabetes or
32  obesity. At the workshop, results from an extensive information survey were evaluated by experts

---

[17]For example, the National Library of Medicine's Gene Expression Omnibus (GEO): a public functional genomics data repository supporting MIAME-compliant data submissions. Array- and sequence-based data are accepted. Tools are provided to help users query and download experiments and curated gene expression profiles.

[18]**Knowledge mining** is the computerized extraction of useful, often previously unknown, information from large databases or data sets using sophisticated data search capabilities and statistical algorithms to discover patterns and correlations and then interpret this new information in the context of systems biology to create new knowledge.

1   on the strength of the associations identified. The effort integrated both traditional and new types
2   of data, including approximately 870 findings from more than 200 human studies; and the most
3   useful and relevant endpoints in experimental animals and *in vitro* assays (e.g., ToxCast™ and
4   Tox21). The environmental factors identified and discussed at the workshop included maternal
5   smoking and nicotine, arsenic, persistent organic pollutants, organotins, phthalates, bisphenol A
6   (BPA), and pesticides. Overall, the workshop results suggest that associations can be made between
7   environmental factors and type 2 diabetes or obesity, but causality was more difficult to assign
8   (Table 5).

**Table 5. Summary of Literature Review Findings and Expert Judgments Concerning Causal Relationships**

| Chemical/ Environmental Factor | Outcome | Association/ Causality | Conclusions from Breakout Group |
|---|---|---|---|
| Maternal smoking and nicotine | Childhood obesity | Association, likely causal | Likely causal supported by epidemiology data and animal studies (Behl et al. 2013). |
| Arsenic | Diabetes | Association | Sufficient support for an association between arsenic and diabetes in populations with relatively high exposure levels (≥ 150 µg arsenic/L in drinking water) (Maull et al. 2012). |
| Organochlorine persistent organic pollutants | Diabetes | Association | Sufficient for a positive association of some organochlorine persistent organic pollutants with type 2 diabetes (Taylor et al. 2013). |
| Organotins | Obesity | Suggestive of an association in animal and *in vitro* models | Current data from human studies of exposure to organotins are nonexistent regarding an association with diabetes or obesity. Recent animal and mechanistic studies report stimulatory effects of tributyl tin on adipocyte differentiation (*in vitro* and *in vivo*) and an increased amount of fat tissue (i.e., larger epididymal fat pads) in adult animals exposed to TBT during fetal life. Although the organotin "obesogen" literature is relatively new, with few studies, the quality of the existing experimental studies was considered high by the breakout group (Thayer et al. 2012). |
| Bisphenol A (BPA) | Diabetes | Suggestive of an association | Overall, this breakout group concluded that the existing data, primarily based on animal and *in vitro* studies, are suggestive of an effect of BPA on glucose homeostasis, insulin release, cellular signaling in pancreatic β cells, and adipogenesis (Thayer et al. 2012). |
| Phthalates | Diabetes or obesity | Insufficient data to assess | Current data from human studies of exposure to phthalates provide insufficient evidence of an association with diabetes or obesity (Thayer et al. 2012). |

## Environment-Wide Association Studies[19]

1   Diabetes varies in the population due to both genetic and environmental factors but understanding
2   these interactions has been difficult. Using an Environment-wide Association Study approach, Patel
3   et al. (2012b) investigated the problem of many possible contributing factors by integrating
4   genomic and toxicological data to obtain a candidate list of interacting genes, genetic variants, and
5   environmental factors associated with type 2 diabetes. The method involved three steps. First,
6   genetic and environmental data were summarized from VARIMED (VARiants Informing MEDicine; a
7   genetic association database) and the National Health and Nutrition Examination Survey (NHANES,
8   an environmental exposure and effects database). VARIMED contains data on 11,977 gene variants,
9   9,752 genes, and 2,053 individuals; NHANES includes 261 genotyped loci, 266 environmental
10  factors measured in blood and urine, and clinical measures for the same individuals. They identified
11  several environmental factors that positively or negatively affected risks for type 2 diabetes,
12  including nutrients and persistent organic pollutants. They reported 18 human genetic variations
13  (SNPs) and 5 serum-based environmental factors that interacted in association with type 2
14  diabetes. Thus Patel et al. (2013, 2012b) successfully identified association linking diabetes, genes,
15  gene variants, and environmental factors.

16  This approach demonstrates a knowledge mining method that can be applied broadly to any
17  number of common diseases to identify possible interactions between genetic and environmental
18  factors and risks of disease. In *Genetic Variability in Molecular Response to Chemical Exposure*, Patel
19  and Cullen (2012) review what has been learned to date with these types of efforts and discuss a
20  more comprehensive representation of chemical exposures as the "envirome" and how we might
21  use it to examine the interplay of genetics and the environment.

## Itemset Associations between Prediabetes/Diabetes and Chemical Exposures

22  We followed up efforts by Thayer et al. (2012) and Patel et al. (2013, 2012b), using two
23  independent frequent itemset mining analyses of the NHANES data. Frequent itemset mining is a
24  data mining approach commonly used in business intelligence to derive marketing and pricing
25  strategies or to identify credit risks. For example, grocery stores use frequent itemset mining to
26  uncover products that are typically purchased together to determine pricing strategies (e.g., a
27  grocer does not want to place items commonly purchased together on sale at the same time and
28  might raise the price of an item commonly purchased with a sale item). Similarly, this technique can
29  be used with the NHANES data to uncover a chemical or group of chemicals that tend to be
30  associated with specific diseases.

31  We focused our analyses on the 2003–2004 NHANES cohort and evaluated associations between
32  diabetes and individual chemicals. We also focused on the 2009–2010 NHANES cohort and
33  evaluated associations among diabetes and a more complex lists of chemicals.[20] Both analyses
34  focused on metals.

---

[19]This section is adapted largely from Patel et al. (2012b) and (2013) with the assistance of Dr. Patel.
[20]Both analyses use the Apriori algorithm (Borgelt 2013) to generate "rules" where X ≥ Y is read "X is associated with outcome Y." Our first study constrained the rule to read "prediabetes/diabetes is associated with chemical Y," or prediabetes/diabetes ≥ chemical Y. Our second study constrained Apriori to return prediabetes/diabetes as the outcome.

1    *Prediabetes/Diabetes and Individual Chemical Exposures* – These results suggest that type 2
2    prediabetes/diabetes is most often associated with lead and cadmium (blood or urine), with a
3    suggestive association with arsenicals. Type 2 prediabetes/diabetes is not associated with cesium
4    and uranium. Table 6 lists the resulting rules[21] showing the association or lack of association
5    between diabetes and all of the metals monitored in NHANES.

6    When interpreting lift,[22] support,[23] and confidence,[24] we believe lift is the most informative to start
7    with, followed by the other measures. If a rule has a lift value close to 1, the rule has little predictive
8    value, regardless of the support and confidence. Once an analyst has identified models that are
9    significantly different from random (lifts > 1), the analyst will typically then examine the support
10   and confidence.

11   Support provides an indication of the percentage of people surveyed by the NHANES program that
12   have both type 2 prediabetes/diabetes (the antecedent) and high blood lead, for instance (11% in
13   this case). The support indicates what proportion of the population might be expected to have type
14   2 prediabetes/diabetes and high blood lead, assuming the NHANES sample is a truly representative
15   sample of the U.S. population (in this case 11%).

16   The confidence tells the analyst how strong the rule is. In other words, confidence tells the analyst
17   the percentage of people with type 2 prediabetes/diabetes (the antecedent) that have high blood
18   lead, for instance (34% in this case). This is equivalent to the prevalence of Type 2
19   prediabetes/diabetes in individuals that have a high level of the particular metal, and is a potential
20   indicator of risk.

---

[21]Ruleset is a collection of one or more rules used, in this case, to predict association between diabetes and chemical exposures (Oracle 2013a).

[22]Lift is a measure of how much better prediction results are using a model than could be obtained by chance (Oracle 2013b). A lift of 1 means the rule is no better at predicting the outcome than random chance. Thus, knowing that someone in this NHANES cohort is defined as prediabetic or diabetic provides a 1.44 times better chance to predict that the person has high blood lead, compared to random. The lift close to 1 provides no better indication of a person's urine uranium or cesium concentration compared to random guessing knowing that they are prediabetic or diabetic.

[23]Support is the percentage of subjects in the entire data set/database that have both the antecedent/condition and the predicted outcome. This can also be thought of as the number of subjects in the entire data set/database where the rule is true.

[24] Confidence is the percentage of the people who meet the antecedent/condition criteria that also meet the outcome criteria. For instance, 34% of the people in this NHANES cohort defined as being either prediabetic or diabetic also have high blood lead.

---

**Table 6. Apriori Rule Associations between Type 2 Prediabetes/Diabetes and Chemical Exposures.**

| Antecedent/Condition | Predicted Outcome | Lift | Support | Confidence | Conclusion |
|---|---|---|---|---|---|
| Type 2 Prediabetes/Diabetes | High blood lead | 1.44 | 0.11 | 0.34 | Association |
| Type 2 Prediabetes/Diabetes | High urine cadmium | 1.43 | 0.13 | 0.43 | Association |
| Type 2 Prediabetes/Diabetes | High blood cadmium | 1.26 | 0.09 | 0.30 | Association |
| Type 2 Prediabetes/Diabetes | High urine arsenobetaine | 1.25 | 0.10 | 0.33 | Association |
| Type 2 Prediabetes/Diabetes | High urine lead | 1.20 | 0.09 | 0.28 | Association |
| Type 2 Prediabetes/Diabetes | High urine total arsenic | 1.18 | 0.09 | 0.31 | Association |
| Type 2 Prediabetes/Diabetes | High blood total mercury | 1.12 | 0.09 | 0.30 | Association |
| Type 2 Prediabetes/Diabetes | High urine cesium | 1.03 | 0.08 | 0.25 | No association |
| Type 2 Prediabetes/Diabetes | High urine uranium | 1.01 | 0.07 | 0.24 | No association |

1  *Prediabetes/Diabetes and Multiple Chemical Exposures* – Table 7 lists the results showing
2  associations between multiple chemicals and prediabetes/diabetes. The rule with the highest lift
3  (1.46 times better than random) is where NHANES subjects had high urine cadmium, high blood
4  lead, and high total urine arsenic. This rule is present in 11% of the 2009–2010 NHANES cohort,
5  suggesting it might be true for 11% of the U.S. population at the time of study, assuming NHANES is
6  a good random sample. Of all the subjects who had high urine cadmium, high blood lead, and high
7  total urine arsenic, 59% also were either prediabetic or diabetic. Not surprisingly, the rule with the
8  next highest lift is the same as the first, except these subjects also had high urine lead levels. This
9  rule had a support of 10% and a confidence of 58%. Overall, this analysis supports strong
10  associations between cadmium, lead, and total urine arsenic and type 2 prediabetes/diabetes due
11  to their frequent occurrence in the top ranked rules. Cesium and cobalt occurred less frequently
12  and would be expected to be less strongly associated.

13  *Synthesis of Frequent Itemset Mining Results* – Lead and cadmium exposure are highly likely to
14  be associated with type 2 prediabetes/diabetes. High lead levels occurred in 9 of 10 and cadmium
15  in 8 of 10 of the top-ranked rules in Burgoon's data set. Further evidence is provided by the results
16  where blood lead, blood cadmium, and urine cadmium were the highest rated outcomes based on
17  lift. Confirmatory evidence exists that these metals also might be elevated in other diabetic
18  populations (Afridi et al. 2008). Low-dose mixtures of lead, cadmium, and arsenic might induce
19  oxidative stress (Fowler et al. 2004), and evidence suggests that cadmium might induce
20  hyperglycemia in rats (Bell, RR et al. 1990).

**Table 7. Apriori Rule Associations between Type 2 Prediabetes/Diabetes and Exposure to Multiple Chemicals**

| Antecedent/Condition | Predicted Outcome | Lift | Support | Confidence | Conclusion |
|---|---|---|---|---|---|
| High urine cadmium<br>High blood lead<br>High total urine arsenic | Type 2 Prediabetes/Diabetes | 1.46 | 0.11 | 0.59 | Association |
| High urine cadmium<br>High urine lead<br>High blood lead<br>High total urine arsenic | Type 2 Prediabetes/Diabetes | 1.44 | 0.10 | 0.58 | Association |
| High urine cadmium<br>Low urine cobalt | Type 2 Prediabetes/Diabetes | 1.40 | 0.11 | 0.56 | Association |
| High urine cadmium<br>High blood lead | Type 2 Prediabetes/Diabetes | 1.38 | 0.17 | 0.56 | Association |
| High urine cadmium<br>High urine lead<br>High blood lead | Type 2 Prediabetes/Diabetes | 1.38 | 0.15 | 0.56 | Association |
| High urine cadmium<br>High urine cesium<br>High blood lead | Type 2 Prediabetes/Diabetes | 1.38 | 0.11 | 0.56 | Association |
| High urine cadmium<br>High blood cadmium<br>High blood lead | Type 2 Prediabetes/Diabetes | 1.37 | 0.13 | 0.55 | Association |
| High urine lead<br>High blood lead<br>High total urine arsenic | Type 2 Prediabetes/Diabetes | 1.37 | 0.12 | 0.55 | Association |
| High urine cesium<br>High blood lead<br>High total urine arsenic | Type 2 Prediabetes/Diabetes | 1.37 | 0.10 | 0.55 | Association |
| High urine cadmium<br>High urine lead<br>High blood cadmium<br>High blood lead | Type 2 Prediabetes/Diabetes | 1.37 | 0.11 | 0.55 | Association |

1   Taking these results together, uranium and cesium are not likely to be associated with type 2
2   prediabetes/diabetes. Whether mercury is likely to be associated with type 2 prediabetes/diabetes
3   remains unclear.

4   Extrapolating these results to the U.S. population suggests that a large proportion (>50%) of the
5   population with elevated lead, cadmium, and arsenic levels might have type 2
6   prediabetes/diabetes. Although these data are not sufficient to demonstrate causality, they do
7   suggest that mixtures of these metals are associated with type 2 prediabetes/diabetes. Possible
8   explanations include (1) the mixture of these chemicals might cause type 2 prediabetes/diabetes;
9   (2) prediabetic/diabetic phenotypes might alter the absorption, distribution, metabolism, and

1   excretion of these metals; or (3) only one of these chemicals might be associated with type 2
2   prediabetes/diabetes, while the absorption, distribution, metabolism, and excretion properties of
3   the other chemicals are impacted by the first. Evidence exists that the three metals work together to
4   induce oxidative stress, and cadmium itself might induce hyperglycemia in rats. These results
5   suggest that further studies should be conducted to ascertain potential causality.

6   Further, these results demonstrate that Frequent Itemset Mining yields fruitful results and
7   hypotheses that can be used to identify associations between chemical body burdens and potential
8   disease endpoints. The results also illustrate ways that data mining methods developed for other
9   fields can be implemented to identify predictive biomarkers of exposure and health outcomes.

### Example: Characterizing Human Susceptibility and Population Variability

10  Risk managers can begin to identify populations with genetic susceptibility to Type 2
11  prediabetes/diabetes in their communities by combining the frequent itemset mining results above
12  with data mining of human genetic variability data, health outcomes, and an understanding of
13  disease processes and chemical MOAs. Combining this information with census demographic data,
14  geographic information systems, and exposure models will further drive the possibilities of
15  pinpointing specific geographic susceptible populations. In this prototype, we identify a potentially
16  susceptible population to Type 2 prediabetes/diabetes by combining the cadmium-disease
17  association, known gene-disease associations, and knowledge of risk allele frequencies in human
18  ethnic groups.

19  Recently, a combination of EWAS and GWAS was performed that examined potential interactions
20  between SNPs (i.e., a mutation of a single nucleotide within the DNA of a gene sequence),
21  environmental chemical levels in blood and urine, and health outcomes—specifically type 2
22  diabetes—using data from NHANES (Patel et al. 2013). Although support for genotype and chemical
23  interactions was limited, interesting interactions were noted between the nonsynonymous coding
24  SNP rs13266634 in the SLC30A8 gene and cis- and trans-beta-carotene and gamma-tocopherol.

25  The SNP rs13266634 has been noted as being associated with type 2 diabetes previously (Rung et
26  al. 2009, Takeuchi et al. 2009, Timpson et al. 2009, Pare et al. 2008, Diabetes Genetics Initiative of
27  Broad Institute of Harvard et al. 2007, Scott et al. 2007, Sladek et al. 2007, Steinthorsdottir et al.
28  2007, Zeggini et al. 2007). SLC30A8 is a zinc transporter found in the pancreatic beta-cell secretory
29  vesicles. Zinc has been associated with insulin biosynthesis (Emdin et al. 1980), and chronic
30  decreased zinc intake has been associated with an increased risk of diabetes (Miao et al. 2013). The
31  risk allele in rs13266634 is C (Sladek et al. 2007), while the minor allele is T (NCBI 2012b).

32  Risk managers might find the genotype and allele frequency data in dbSNP to be helpful in
33  understanding population variance and to help identify susceptible populations. For instance, from
34  a random sample of 100 individuals of Mexican descent in Los Angeles, 66% were homozygous for
35  the risk allele, 30% were heterozygous, and 4% were homozygous for the nonrisk allele (NCBI
36  2012b). If we assume the sampling is representative of the entire population of Mexican-descended
37  residents of Los Angeles, then approximately 66% of these individuals might be at an increased risk
38  of developing diabetes, independent of their body mass index (OMIM 2012). Heterozygous
39  individuals (30% of the population) might also carry some risk and might be more affected by their
40  zinc intake (i.e., increased zinc intake might be protective). Likewise, the heterozygous individuals
41  might be more sensitive to other metals, chemicals, or dietary factors that might compete with zinc
42  for absorption, or they might be more sensitive to chemicals that could interfere with zinc

1  metabolism, transport, and insulin biosynthesis. Given the high rate of zinc deficiency in Mexican
2  children that is not correlated with socioeconomic status, finding zinc deficiency in children of
3  Mexican descent living in Los Angeles might not be surprising, especially if diet plays a significant
4  role in the deficiency (Morales-Ruan Mdel et al. 2012).

5  Cadmium exposure will be of concern to individuals who are homozygous or heterozygous for the
6  risk allele. Cadmium has been shown to compete with zinc transporters and might lead to beta-cell
7  dysfunction, lack of insulin production, and ultimately hyperglycemia (El Muayed et al. 2012).
8  Individuals with the rs13266634 risk allele could be more sensitive to cadmium exposures than the
9  rest of the population.

10  Through database mining and an understanding of the allele disease pathway and a chemical's
11  adverse outcome pathway, we can identify potentially susceptible populations more easily. This
12  example could be extended by examining cadmium exposure data for the Los Angeles area and
13  using a geographic information systems approach with census data to identify potentially
14  susceptible individuals, based on the allele probabilities. This type of predictive modeling could
15  help advance more targeted community-level responses in the future.

### 3.2.2. Short-Term *In Vivo* Models – Alternative Species

16  Alternative species (i.e., nonmammalian species) provide *in vivo* models for identifying hazards,
17  integrating dose-response effects, and understanding pathways and apical effects useful for
18  assessing chemical risks to humans and to other species. The shorter life spans of alternative
19  species enable the evaluation of toxicity over the full life span of the intact organism, facilitating
20  study of the entire etiology of disease from the MIE to apical outcomes, including more complex
21  phenomena such as birth defects or neurobehavioral impairment.

22  Alternative species studies are playing a progressively more integral role in chemical testing,
23  hazard identification, and dose-response assessment for both human and nonhuman species (ECHA
24  2013b, Perkins et al. 2013, EPA 2012d, EC 2011, Schug et al. 2011, OECD 2004). Both the European
25  Chemicals Agency (ECHA) and EPA use alternative species tests as part of required tests for
26  endocrine disruptors (EPA 2012e, 2009a). Alternative species studies can be used for prioritization
27  and screening or as the basis for Tier 2 type assessments.[25]

### Tier 2 Prototype: Using Alternative Species to Identify Thyroid Hormone Disruption

28  Endocrine disruptors are chemicals that interfere with the body's endocrine system and produce
29  adverse effects in both humans and wildlife. In a state-of-the-science review, the World Health
30  Organization (WHO) concluded that thyroid disruption-associated neurobehavioral disorders are

---

[25]The types of alternative or nonmammalian species can vary widely. Considerable work in toxicology has been done with fish, but work in very simple organisms such as yeast also provides insight into cellular regulation at multiple levels that control core biological processes and enable cells to respond to genetic and environmental changes (Yeung et al. 2011). Zhu et al. developed a network reconstruction approach that simultaneously integrates different types of data, and constructs a probabilistic causal network representing complex cell regulation: endogenous metabolite concentration, RNA expression, DNA variation, DNA–protein binding, protein–metabolite interaction, and protein–protein interaction data. Causal regulators of the resulting network were identified and provide insight into the mechanisms by which variations in network interactions affect gene expression and metabolite concentrations (Zhu et al. 2012).

1   occurring in children, and the incidence of these disorders has increased in recent decades (WHO
2   2012). Normal thyroid function is essential for normal brain development, particularly during
3   pregnancy and after birth. Additionally, thyroid hormones are crucial to inner ear and bone
4   development, and bone remodeling and physiological functions such as metabolism (De Coster and
5   van Larebeke 2012). Internationally agreed-upon and validated test methods for identification of
6   endocrine disruptors capture a limited range of the known endocrine disrupting effects (Miller, MD
7   et al. 2009). In its state-of-the-science review, WHO advised that existing testing protocols do not
8   characterize completely all essential functions and that adverse effects "are being overlooked"
9   (WHO 2012). Identifying environmental factors that might disrupt normal thyroid function and
10  impact public environmental health is needed, given the key role that thyroid hormone plays for
11  normal development and physiologic functions in all vertebrates (Woodruff and Sutton 2011,
12  Miller, MD et al. 2009).

13  In regulating development, the role of the thyroid hormone is of particular toxicological interest
14  because thyroid hormone-dependent post-embryonic development is a common feature of
15  vertebrate ontogeny (Paris and Laudet 2008). This period of development is typically characterized
16  by transient elevations of thyroid hormone that elicit species-specific physiological and
17  morphogenetic responses with lasting developmental consequences. Transitions from tadpoles to
18  juvenile frogs and body plan reorganization in flatfish are two nonmammalian examples of thyroid
19  hormone-controlled events. Human and vertebrate post-embryonic neurodevelopment is thyroid
20  hormone-dependent and deviations from normal thyroid hormone concentrations at critical times
21  are associated with a variety of neurological defects and deficits (Zoeller et al. 2002). The timing (or
22  window) of exposure is critical as the impact of thyroid hormones changes as the brain develops
23  (Zoeller and Rovet 2004). Thyroid hormone regulation is generally essential for normal
24  development in vertebrates, thereby establishing the basis for cross-species extrapolation of
25  developmental risks. Several methods using alternative species have been proposed to measure
26  these outcomes for thyroid pathways (Makris et al. 2011, Nichols et al. 2011).

27  A key factor in thyroid hormone related risk assessment is the ability to examine hormone
28  disruption and the resultant developmental disruption at higher levels of tissue organization.
29  Results from omics technologies and other thyroid hormone toxicity assessments, such as EPA's
30  ToxCast™ chemical screening efforts (EPA 2008), can be linked to adverse outcome data from
31  alternative species studies. Two examples are:

32      1.  The construction of regulatory networks using time series data in genotyped populations
33          and integration of multiple data types (i.e., endogenous metabolite concentrations, RNA
34          expression, DNA variation, DNA-protein binding).

35      2.  If a chemical is identified as a potential developmental disruptor in HTS, more information
36          on *in vivo* effects might be required to establish dose-response relationships, windows of
37          susceptibility, potential impacts of maternal exposure on progeny, and existence of subtle
38          impacts on behavior, learning, and memory.

Systems and Pathway Models

39  As discussed throughout this document, understanding of systems biology and the events leading to
40  an adverse effect are central features for the use of molecular biology data in risk assessment.
41  Pathway analyses are useful to inform extrapolation across species and to aid in characterizing the
42  variability within populations through identifying and describing both MIEs and key biological

1    events leading to adverse outcomes. They can also help identify how human-focused screening data
2    can inform ecological risk assessment. Although making quantitative predictions of disease risks
3    based on today's system biology or adverse outcome models is often very difficult, pathway
4    assessments are critical to advancing dose-response assessment.

5    Figure 19 illustrates an example for thyroid hormone disruption. Disruption of the thyroid
6    pathways can occur by thyroid peroxidase inhibition, iodine uptake (sodium iodide symporter)
7    inhibition, enhanced phase II metabolism (glucuronosyltransferases or sulfotransferases) via
8    alterations in specific genes (CAR/PRX [constitutive androstane receptor/prename x receptor]) or
9    receptors (AhR), enhanced cellular transport, deiodinase inhibition, and interference with thyroid
10   receptor function. In humans, this leads to birth defects, decreased IQ, and metabolic disorders. In
11   rats, increased TSH leads to thyroid hyperplasia and cancer.

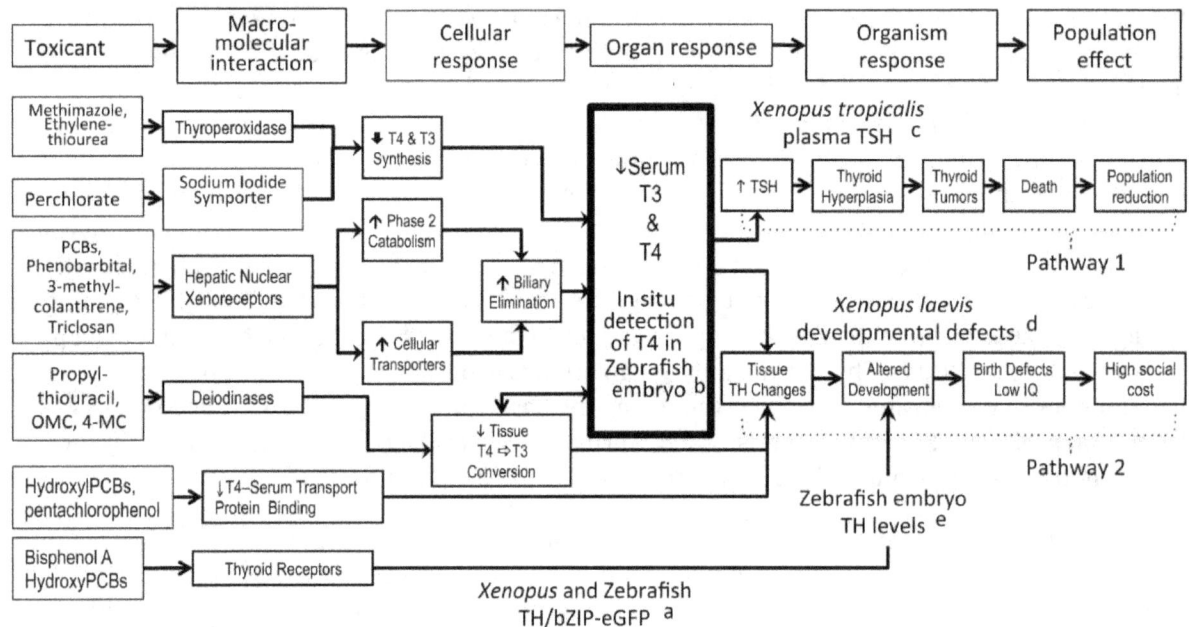

Figure 19. Major pathways involved in thyroid disruption with example toxicants and alternative models applicable to both human and ecological hazard assessment (Perkins et al. 2013). Reproduced with permission from *Environmental Health Perspectives.*[26]

---

[26] The thick black outlined box indicates the critical event of serum level concentrations of thyroid hormones. Pathway 1: rat pathway leading to tumors via thyroid hyperplasia. Pathway 2: principle pathway of concern affecting humans. Abbreviations: IQ, intelligence quotient; 4-MC, 4-methylbenzylidene camphor; OMC, octyl methoxycinnamate; $T_3$, triiodothyronine; $T_4$, thyroxine; TR, thyroid receptor. [a]Quantification of plasma TSH levels in *Xenopus tropicalis.* [b]Direct quantification of intrafollicular concentrations of $T_4$ in zebrafish embryos. [c]Detection of developmental defects with *X. laevis* metamorphosis assay. [d]Detection of developmental defects using zebrafish embryos. [e]Reporter gene (eGFP) detection of TR activity.

*Informing Dose-Response Assessment*

1  Understanding causal mechanisms and their relationships to adverse outcomes provides insights
2  into both hazard identification and dose-response assessment. Although quantitatively predicting
3  human disease risks based on knowledge of causal mechanisms is currently difficult, several
4  approaches using alternative species data provide information on the potency of chemicals that
5  cause effects: biomarkers of exposure and effect, relative potency to induce adverse effects, and
6  understanding of the complexities of dose-response relationships.

*Biomarkers of Exposure and Effects*

7  Key events in the perturbed pathway can be represented with biomarkers of exposure and effect. In
8  situations where considerable systems biology information links the event to outcomes, a
9  biomarker might provide a measure of hazard for risk assessment. In the thyroid disruption
10  example, upstream events converge on serum levels of the thyroid hormones, triiodothyronine (T3)
11  and thyroxine (T4), and downstream events occur in peripheral tissues where a significant degree
12  of species-specific effects are seen (Figure 20). As a result, serum T4 levels can be used as a
13  biomarker of thyroid function across species. In the laboratory, researchers use T4 and thyroid
14  stimulating hormone levels in fish and frogs to assess the thyroid disrupting potential of chemicals
15  (Tietge et al. 2013, Thienpont et al. 2011). To assess human exposures, the Centers for Disease
16  Control and Prevention (CDC) has used decreased serum levels of T4 (noted as key event in Figure
17  20) and increased levels of TSH measured in the U.S. population to infer increased potential risks
18  for thyroid dysfunction-related disorders at low levels of perchlorate exposures (Lau et al. 2013,
19  Blount et al. 2007).

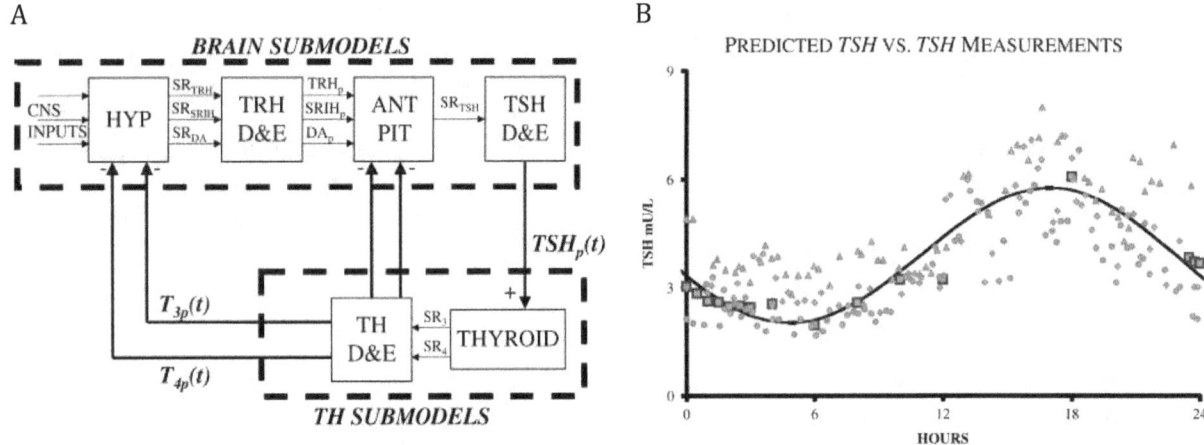

Figure 20. Dose-response relationships. Within species, significant advances are being made in quantitative systems biology modeling (Eisenberg et al. 2008). Panel A: Overall feedback control system model of thyroid hormone regulation with three source organ blocks (hypothalamus [HYP], anterior pituitary [ANT PIT], and thyroid glands [THYROID]); three sink blocks—for TRH, TSH, and T3 and T4 distribution; and elimination (elimination = metabolism and excretion) (D&E). TRH = thyrotropin-releasing hormone; TSH = thyroid-stimulating hormone; T3 = triiodothyronine; T4 = thyroxine; SR = secretion rate; p = plasma or portal plasma for TRH-related components; DA = dopamine; SRIH = somatostatin. Panel B: Feedback control system (FBCS) model validation study results. Predicted normal circadian TSH versus independent TSH data (not used in fitting the FBCS model) (triangles and diamonds represent data from Sarapura et al. (2002), circles represent data from Samuels et al. (1994). Also shown (squares) are the mean TSH data from the larger database used to fit the FBCS model of Blakesley et al. (2004). Reproduced with permission from Mary Ann Liebert, Inc.

### Relative Potency

1    Identification of pathways and assays impacted by chemicals of known toxicity can be useful in
2    initial prioritization of many compounds. These can be identified through development of
3    predictive models built on relationships between *in vitro* ToxCast™ assay results and *in vivo* effects
4    caused by known developmental toxicants (Sipes et al. 2011). A chemical's potency can be further
5    refined using focused *in vivo* tests with alternative species to provide informative dose-response
6    data and exposure window relationships. Alternative species provide easily manipulated model
7    systems that can detect effects caused by mechanisms not assessed by *in vitro* HTS. For example,
8    zebrafish were used as a screening model to assess the 309 EPA ToxCast™ Phase I chemicals for
9    developmental toxicity to both humans and ecological species. In fish embryos or larvae, 191 (62%)
10   chemicals were toxic (death or malformations) to the developing zebrafish. Both toxicity incidence
11   and potency were correlated with chemical class and hydrophobicity. As an integrated model of the
12   developing vertebrate, the zebrafish embryo screen provides information relative to overt and
13   organismal toxicity. In 12 classes of chemicals, 100% of the chemicals induced developmental
14   toxicity, 4 classes of which induced developmental toxicity with an average concentration at 50% of
15   the maximum level ($AC_{50}$)[27] below 4 μM. Translating such results directly into a dose-response for
16   human risks is difficult, but results of Padilla et al. (2012) show that alternative species can be used

---

[27] In high-throughput screening (HTS) assay, $AC_{50}$ is the concentration at which an assay is inhibited or activated by 50% when compared to control values. This value is useful in comparing assay results.

1    to build relative rankings of chemicals based on their potency to cause adverse effect. Such rankings
2    can be used for ranking and prioritizing chemicals or classes of chemicals for additional evaluation.

## Dose-Response Relationships

3    Chemical dose-response relationships characterized in one alternative species might be
4    extrapolated to other species or to humans, using a pathway-based approach (Perkins et al. 2013).
5    Because many biological functions and disease pathways are conserved across species, similarity of
6    genes encoding those pathways can support direct comparisons of pathway or genomic effects
7    between species. Where pathways are highly conserved between species, this information can be
8    used to extrapolate dose-response relationships in alternative species to analogous relationships in
9    mammals. For example, pathways in the hypothalamus-pituitary-gonad axis are highly conserved
10   among vertebrates, which been used to show that chemical effects in fathead minnows are
11   predictive of endocrine disrupting effects in rats (Ankley, G. T. and Gray 2013).

12   Pathway effects defined through gene expression changes can be used to define a benchmark dose
13   or sensitivity of an animal to a chemical (Thomas, RS et al. 2011). Benchmark concentrations
14   derived from aqueous exposures of alternative species can be translated to oral equivalents in
15   other species, such as humans, by applying a dose scaling factor composed of a simple reverse
16   toxicokinetics approach to estimate the blood dose and amount of metabolism in the target species
17   (Wetmore et al. 2012). Using this approach, chemical concentration effects can be translated from
18   alternative species to mammalian species. See Figure 20 for an example of how systems biology can
19   inform dose-response extrapolation within species.

20   However, this type of an approach has added uncertainty, and may generally increase uncertainty
21   to an unacceptable level, which precludes the calculation of a reference value, including a
22   provisional value. There is uncertainty with respect to defining a benchmark dose based on gene
23   expression changes and with respect to the pathway context and interpretation. For instance,
24   changes in gene expression do not directly translate into changes in protein expression or activity.
25   In addition, it is well known that signaling and metabolic pathways within the cell are intersecting
26   and inter-related. There is uncertainty with respect to the dose-response changes at particular key
27   events and how downstream key events may be altered by other intervening pathways. Thus,
28   calculating a benchmark dose for a pathway itself is fraught with challenges and additional
29   uncertainty that current reference value approaches do not take into account. In all likelihood,
30   accounting for these additional sources of uncertainty may require new uncertainty factors to be
31   developed, and increases the likelihood that an unacceptable level of uncertainty may be
32   encountered.

33   Thus, it is more likely that, until better, less uncertain methods and techniques are developed and
34   used, pathway-based effects based on gene expression are more suitable for screening level
35   decisions and less suitable for reference value derivation.

## Species to Species Extrapolation

36   For most species, qualitative predictions are likely to be tenable based on hypothalamus-pituitary-
37   thyroid (HPT) dependent pathways, that is, iodine uptake. Altered iodine uptake hinders
38   development, but the most sensitive outcome indicator might be different. In rats, thyroid hormone
39   disruption can lead to thyroid tumor development (Hurley 1998), while in frogs, metamorphosis is

1   disrupted (Degitz et al. 2005). Quantitative predictions might not be feasible for many species due
2   to limited data on downstream endpoints and key events (Perkins et al. 2013) .

3   Normal thyroid hormone-dependent post-embryonic development requires coordinated spatial-
4   temporal control of thyroid hormone activity. Such activity is regulated not only through the
5   classical features of the HPT axis, but also through peripheral mechanisms external to the
6   hypothalamus, pituitary, and thyroid tissues, such as differential regulation of deiodinase activity,
7   hepatic metabolism and excretion of thyroid hormones, thyroid hormone receptor regulation, and
8   transmembrane thyroid hormone transport. Of these major controlling processes, the mechanisms
9   of the central HPT axis are considered to be generally conserved across vertebrate species and
10  useful for comparative efforts; however, those of the peripheral tissues are typically more divergent
11  and must be used with care in cross-species analysis.

## Population Variability

12  Understanding the variation of an individual relative to population variation can be key to
13  identifying an adverse effect on a population. Polymorphisms affecting drug responses can vary
14  widely in populations. In humans, 20–25% of prescription drugs are metabolized in the liver by
15  cytochrome P450 CYP2D6 where variants confer widely different rates of drug metabolism, such
16  that some people might respond with an onset of toxicity while others fail to experience efficacy
17  (Ingelman-Sundberg 2005). Variants causing unanticipated results can comprise a significant
18  portion of a population and that distribution can vary widely across populations (Sistonen et al.
19  2007, Ingelman-Sundberg 2005, Andersen et al. 2002, Wooding et al. 2002). Understanding the
20  variation in adverse responses across a diverse testing population helps reduce the uncertainty of
21  extrapolating laboratory data to real populations. Differential response to chemicals is an important
22  consideration in ecological risk assessment where not only potentially sensitive subpopulations
23  might exist, but also sensitive species.

24  Approaches have been developed to incorporate population diversity into toxicity testing through
25  the use of large collections of different genetic lines of mice or cell cultures derived from them
26  (O'Shea et al. 2011, Rusyn et al. 2010, Harrill et al. 2009). Alternative species could be especially
27  useful for incorporating population variability into toxicity testing. The diversity in laboratory lines
28  and outbred populations of fish can be high, especially if populations are collected from different
29  areas impacted by pollutants (Williams and Oleksiak 2011, Guryev et al. 2006). Divergent lines of
30  zebrafish can be used to examine variation in responses to chemicals in addition to determining
31  possible genetic factors influencing adverse effects. Using this approach, Waits and Nebert (2011)
32  crossed zebrafish lines displaying different levels of sensitivity to developmental cardiotoxicity
33  caused by 3,3',4,4',5-pentachlorobiphenyl. The crosses were used in genome-wide quantitative trait
34  loci mapping to identify several genes in addition to the AhR that contribute to the gene-gene and
35  gene-environment interactions that drive developmental toxicity of dioxins and dioxin-like
36  chemicals.

37  Although genetic diversity can be incorporated into testing using a panel of genetically inbred lines,
38  unexpected results can occur. In a study comparing the responses of 19 inbred to 20 outbred
39  zebrafish lines, Brown et al. (2011) found that effects of the endocrine disrupting chemical
40  clotrimazole were dramatically different. Clotrimazole acts by inhibiting P450 activities involved in
41  steroidogenesis production in fish. In inbred fish lines, 11-ketotestosterone production via
42  steroidogenesis was significantly inhibited. In contrast, outbred lines responded with Leydig cell
43  proliferation in testes and normal plasma concentrations of 11-ketotestosterone indicating that the

1     outbred lines could compensate for inhibition by clotrimazole. Here, inbreeding had a strong
2     impact on the diversity and type of response to the endocrine disruptor. Ultimately, the
3     combination of *in vivo* and *in vitro* data should enable development of a weight-of-evidence case as
4     to the toxicity caused by the chemical and whether potential human health effects are likely.

## Cumulative Risks

5     As has been described elsewhere in this document, correct identification of causal perturbations
6     that lead to adverse outcomes will enable determination of which environmental factors are likely
7     to contribute to the cumulative risk for specific outcomes and which are not. Additionally, testing of
8     combinations of chemicals can perhaps be conducted most efficiently in alternative species. For
9     example, alterations in neurosensory functions and intrafollicular thyroxine content of zebrafish
10    exposed to potential disruptors have proven to be useful tools for evaluating multiple chemicals
11    (Raldua et al. 2012, Thienpont et al. 2011, Froehlicher et al. 2009), as has the zebrafish
12    developmental assay (Padilla et al. 2012).

## 3.2.3. Short-Term *In Vivo* Models – Mammalian Species

13    The use of new short-term *in vivo* exposure mammalian bioassays to support Tier 2 assessments
14    are described here. The prototype is drawn from research described in papers by Thomas R.S. et al.
15    (2011)and discussed further in Thomas R.S. et al (2013a, 2013b). The goal of this research was to
16    describe what would be required for the application of short-term *in vivo* transcriptomic assays in
17    predicting chemical toxicity.

## Hazard Identification

18    Short-term *in vivo* transcriptomic assays provide the metabolic capability and systems-level
19    integration of whole animal studies with a more rapid assessment of response to chemical
20    treatment based on molecular-level data. As more data from short-term *in vivo* transcriptomic
21    studies become publicly available, as study designs become standardized, and as gene expression
22    patterns and network perturbations are identified, the ability to predict chemical toxicity
23    comparable to longer term assays is expected to
24    increase. See Text Box 8 for more about the
25    transcriptome.

**Box 8. What is the Transcriptome?**

Ribonucleic acid (RNA) is the functional outcome of deoxyribonucleic acid (DNA) transcription by transcription factors. Researchers can study the transcriptome—the set of all RNA molecules in a given cell—and determine gene expression patterns, or signatures. Specifically, short-term transcriptomic assays in mammalian and alternative species enable observations of the effects of chemical exposure across multiple tissues.

26    For hazard identification, a host of previous studies
27    has demonstrated that transcriptomic signatures
28    from short-term *in vivo* studies can be used to predict
29    both subchronic and chronic toxic responses. A
30    transcriptomic "signature" is typically defined as a
31    subset of genes for which the qualitative or
32    quantitative expression pattern can be used to predict
33    an *in vivo* adverse response with a defined accuracy.

34    To develop a broad-based repertoire of gene expression signatures for hazard prediction, several
35    factors should be considered. First, the number of endpoints included should be sufficient to allow a
36    comprehensive prediction of toxicological hazard. Previous studies that have used gene expression
37    microarray analysis following short-term exposures of chemicals have been limited in the breadth
38    of endpoints examined. These endpoints include the prediction of rat liver tumors (Fielden et al.

1    2011, Uehara et al. 2011, Auerbach et al. 2010, Ellinger-Ziegelbauer et al. 2008, Fielden et al. 2008,
2    Fielden et al. 2007, Nie et al. 2006), mouse lung tumors (Thomas, RS et al. 2009), and rat renal
3    tubular toxicity (Fielden et al. 2005). One strategy that could be employed would be the selection of
4    a battery of tissues, which would include those most frequently positive in rodent cancer bioassays
5    (i.e., liver, lung, mammary gland, stomach, vascular system, kidney, hematopoietic system, and
6    urinary bladder) and tissues commonly affected by noncancer disease. In a previous analysis, these
7    eight tissues cover 92 and 82% of all mouse and rat carcinogens, respectively (Gold et al. 2001).
8    Additional tissues also would need to be added for developmental and reproductive effects, which
9    could include the developing fetus and gonadal tissue.

10   Second, the number of positive and negative chemicals for each endpoint in the studies would need
11   to be sufficient, and the chemical diversity must match the diversity in the chemicals that will
12   ultimately be predicted. For complex toxicological responses such as tumor formation, a previous
13   study estimated that at least 25 chemicals were necessary (Thomas, RS et al. 2009). Third, selection
14   of the time point to perform the gene expression analysis is also a consideration. The time point
15   selection is a balance between cost (i.e., the shorter the time point, the less expensive the study)
16   and a more stable gene expression signature. Among the previous efforts, certain studies relied on
17   much shorter time points (e.g., 5 days), but tended to increase the dose beyond that which would be
18   tolerated in a chronic bioassay (Fielden et al. 2007). Other studies used the same doses as those in
19   the chronic bioassay, but used exposures longer than 5 days (Thomas, RS et al. 2009). In one study
20   that examined the effect of exposure duration, the overall conclusion was that increasing exposure
21   duration increased the predictive performance of the gene expression signatures for genotoxicants
22   (Auerbach et al. 2010).

### Exposure/Dose-Response Assessment

23   As described by Thomas R.S. et al. (2013b, 2012, 2011, 2007), short-term *in vivo* transcriptomic
24   assays have also been applied to dose-response assessment. In a NexGen collaborative effort
25   between EPA and the Hamner Institute, female B6C3F1 mice were exposed to multiple
26   concentrations of five chemicals that were positive for lung or liver tumor formation in a two-year
27   rodent cancer bioassay (Thomas, RS et al. 2012, Thomas, RS et al. 2011). Histological and organ
28   weight changes were evaluated and gene expression microarray analysis was performed on the
29   liver or lung tissues. The histological changes, organ weight changes, and tumor incidences in
30   traditional bioassays were analyzed using standard dose-response modeling methods to identify
31   noncancer and cancer points-of-departure. The dose-related changes in gene expression were also
32   analyzed using a modification of EPA's benchmark dose (BMD) approach (EPA 1995). The gene
33   expression changes were grouped based on both biological processes and canonical signaling
34   pathways. A comparison of the transcriptional BMD values with those for the traditional noncancer
35   and cancer apical endpoints showed a high degree of correlation for specific biological processes
36   (Thomas, RS et al. 2011) and signaling pathways (Thomas, RS et al. 2012). In addition,
37   transcriptional changes in the most sensitive pathway were also highly correlated with the apical
38   responses (see Figure 21). Further studies have also demonstrated the stability of the correlation
39   between transcriptional changes and apical responses across different exposure periods (5 days to
40   13 weeks) (Thomas, RS et al. 2013b). Understanding of the effect of exposure duration on outcomes
41   is a key issue in the design and use of new types of bioassays.

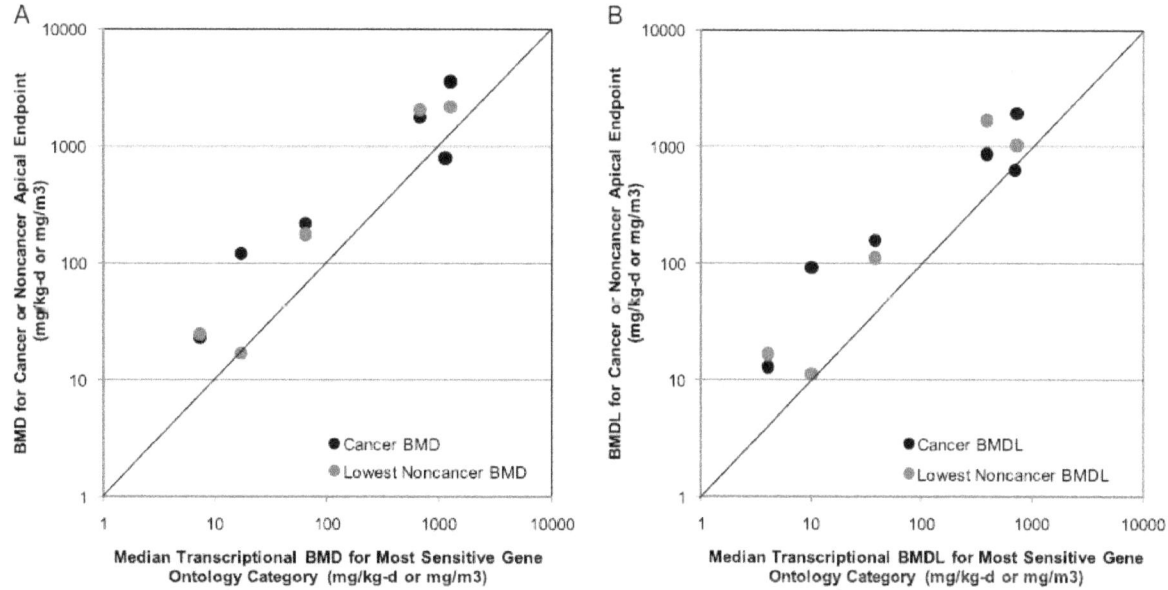

Figure 21. Scatter plot of the relationship between (A) benchmark dose (BMD) and (B) benchmark dose lower limit (BMDL) values for the cancer and noncancer apical endpoints and the transcriptional BMD and BMDL values for the most sensitive GO category. For each chemical and tissue, the BMD and BMDL values for tumor incidence and the lowest noncancer BMD and BMDL values were plotted. For MECL in the lung, no noncancer BMD or BMDL values were plotted because of the absence of histological changes (Thomas, RS et al. 2011). Reproduced with permission from Oxford Journals.

1    With the advent of quantitative high-throughput screening (qHTS), the potential to screen
2    thousands of chemicals for biological activity presents as many challenges as promises. If qHTS can
3    decrease the number of chemicals of interest by 90% (a 10% hit rate across chemicals and assays),
4    this would still overwhelm the throughput of the traditional toxicity testing paradigm. Clearly, a
5    multi-tiered approach to prioritization can lead to more effective applications of animal toxicity
6    testing. As part of this tiered approach, short-term *in vivo* transcriptomic assays provide a tool that
7    incorporates both metabolism and systems-level integration in response to chemical treatment. See
8    also a description of cost savings in Text Box 9. The development of predictive gene expression
9    signatures and dose-response studies would provide a relatively efficient and cost-effective method
10   for both identifying chemicals of concern and estimating a point-of-departure for adverse
11   responses. This information would help support large-scale prioritization and regulatory efforts in
12   the United States and Europe. The gene expression data combined with other data types (e.g.,
13   toxicity data from similar chemicals, PK data) could provide sufficient information to replace the
14   present chronic toxicity and carcinogenicity studies. It should be noted that expression changes can
15   vary depending on dose, time, species, tissue life stage, and individual genetic profile; thus,
16   increasing the complexity of identifying causal relationships between exposure, specific signatures,
17   and outcomes.

## 3.3. Tier 1: Screening and Prioritization

1     This section summarizes new approaches that are available to develop data for screening and
2     prioritizing large numbers of chemicals (i.e., greater than tens of thousands of chemicals) for more
3     focused testing. The increasing maturity of these new approaches has led EPA and other
4     organizations to plan on using these Tier 1 data to prioritize and screen chemicals for immediate
5     regulatory decision, for further testing in Tiers 2 and 3, or in some cases, to add to the weight of
6     evidence in Tier 2 and 3 assessments, especially with respect to identifying pathway or molecular
7     signatures associated with chemical-induced disease.

8     Tier 1 risk assessments are based on *in vitro* assays (including use of human cells), statistical and
9     systems models that focus on molecular molecular targets, QSAR models, and pathways considered
10    relevant to adverse effects or clinical disease. One scientific rationale for using *in vitro* assays is that
11    they probe key events or MIEs that can lead to adverse outcomes. Assay endpoints are designed to
12    represent MIEs and predict subsequent adverse outcomes based on previous studies, both *in vitro*
13    and *in vivo*. The analyses provide the anchoring information critical to characterizing relevance of a
14    "hit" in an *in vitro* assay. Documenting the linkage from assay endpoint to MIE to potential for
15    adversity is thus key to evaluating the relevance of each assay that might be used as part of a Tier 1
16    risk assessment. The evidence for this linkage can come from statistical modeling using *in vivo* and
17    *in vitro* data on the same chemicals, or from detailed biological modeling (e.g., virtual tissue (VT)
18    models or other types of systems biology models).

19    The modeling techniques used in Tier 1 (e.g., QSAR and HTS methods) are designed to assess
20    hundreds to thousands of chemicals in parallel (see Figure 23). In addition to using high-
21    throughput (HT) assays to generate hazard information, moderate- to HT toxicokinetics approaches
22    (here called reverse toxicokinetics or RTK (Rotroff et al. 2010)) are also developed and applied.
23    New approaches can now estimate doses that can activate particular relevant pathways in humans,
24    using data from *in vitro* assays (Wetmore et al. 2012). Bayesian-based exposure models can also be
25    used to generate exposure estimates for chemicals based on production volume and use patterns
26    (Wambaugh and Shah 2010).

27    The data generated from the Tier 1 assays can be used to prioritize chemicals for further study or
28    can simply augment the weight of evidence for chemicals that are already being considered in Tiers
29    2 or 3. For prioritization, from a large set of chemicals for which HTS data are available, one might

1   identify the subset that is likely to interact with known relevant pathways, or which demonstrate
2   pathway disruptions similar to known diseases. Doses at which pathways may be activated or
3   perturbed may be estimated. These estimates may be combined with exposure, occurrence, and
4   other information to select chemicals which may advance into Tiers 2 or 3. Tier 1 data might also
5   directly augment Tiers 2 and 3 weight-of-evidence determinations helping to identify or further
6   characterize pathways alterations associated with disease for sensitive endpoints observed in
7   higher level *in vivo* testing, providing good examples of the integration of the bottom-up and top-
8   down perspectives advocated in the NexGen framework strategy. *In vitro* and modeling data can
9   also be used to guide a next round of *in vivo* data generation.

10  Table 8 provides a brief description and critical review of the tools, methods, and models that could
11  be used in Tier 1.

**Table 8. Summary of Tier 1 NexGen Approaches, Including Weight of Evidence, Pros, and Cons**

| | Tier 1: Screening and Prioritization Categories of Approaches Considered | |
|---|---|---|
| | New QSAR Models | Validated High-Throughput In Vitro Assays |
| Approach: | Uses structural characteristics and experimental data from chemical analogues to predict modes of action, metabolism, hazard, and fate and potency for data-poor chemicals. | Experimentally measures dose-dependent, chemically-induced alterations in biological functions using a range of specific and sensitive *in vitro* assays. Infer potential adverse outcomes based on existing knowledge of other chemical and potential importance of selected biological processes. |
| Weight of evidence: | Determined by quality and amount of existing data, but generally suggestive. | Determined by supporting traditional data and systems biology knowledge, but generally suggestive to likely. |
| Pros: | Data are readily and inexpensively available for all chemicals. If the basis for the QSAR model(s) matches the physical chemistry of the evaluated chemicals, the model(s) generally predicted potency within a factor of 100. Harmonized international approaches are available. | Rapid, inexpensive, multiple bioassay options are available. False negatives and positives for ToxCast™ evaluated assays are low (when testing directly acting chemicals, not toxic metabolites). |
| Cons: | If models do not match the physical chemistry of evaluated chemicals, results are unreliable. Models do not predict active metabolites. | Assay coverage of all important biological processes currently incomplete resulting in false negatives for chemicals that perturbed those processes. Similarly, disorders related to interactions among cell types or tissues cannot be evaluated, that is, reproductive/developmental effects. Limited ability to test for active metabolites or volatiles. False negative rates are of concern. Links to disease outcomes are variable. |

*This document is a draft for review purposes only and does not constitute Agency policy. Do not cite or quote.*

September 2013                                               66

### 3.3.1. QSAR and High-Throughput Virtual Molecular Docking (HTVMD)

1  (Q)SAR[28] models are regression or pattern recognition models that are used in risk assessment to
2  classify or predict the potency of chemicals for toxicological activity, exposure potential, and the
3  like as a function of one or more chemical descriptors. The descriptors are generally the inherent
4  physiochemical properties of the chemical such as atomic composition, structure, substructures,
5  hydrophobicity, surface area charge, and molecular volume. QSAR models require only the inherent
6  properties of the 2-D or 3-D chemical structure as input parameters, and are thus considerably less
7  costly and faster than hazard animal test results. With a variety of QSAR models to choose from
8  (Hansen et al. 2011), and each model having a set of assumptions and a chemical domain of
9  applicability, interpreting QSAR results for use in hazard and dose-response assessment requires
10  expertise.

11  QSAR models have been most commonly used in classification of chemicals with unknown hazard
12  or exposure potential by comparing the "query" chemical's inherent properties with similar
13  properties for a set of chemicals that have known toxicological or exposure potential called the
14  "training set" (Venkatapathy and Wang 2013, EPA 2012c, Goldsmith et al. 2012, OECD 2012, Wang,
15  N et al. 2012b, EC 2010). SAR models provide a qualitative identification of specific hazards (e.g.,
16  suspected carcinogens, mutagens, and reprotoxicants). The commercially available TOPKAT model
17  (TOPKAT, 2013) provides quantitative estimates that can be used to rank chemicals for potency
18  (Venkatapathy and Wang 2013, Venkatapathy et al. 2004). In the European Community, QSAR
19  results are used to prioritize chemicals for additional toxicity testing.

20  At EPA, (Q)SAR models are being used to screen, rank, and categorize chemicals for level of concern
21  in a variety of EPA programs, including Superfund mitigation; the Office of Chemical Safety and
22  Pollution Prevention (OCSPP) High Production Volume Challenge Program and Pre-Manufacture
23  Notice review process; the OCSPP/Office of Water Endocrine Disruptors Screening Program (Weiss
24  et al. 2012); and the Office of Water Candidate Contaminant List. The QSAR models used by EPA
25  include the Sustainable Futures Initiative suite of models, the Organization of Economic Co-
26  operation and Development (OECD) QSAR toolbox models (OECD 2012, 2004), High-throughput
27  Virtual Molecular Docking (HTVMD) (Rabinowitz et al. 2008), MetaCore (Teschendorff and
28  Widschwendter 2012, van Leeuwen et al. 2011), and the TOPKAT model (Rakyan et al. 2011,
29  Venkatapathy et al. 2004).

30  HTVMD models use a ligand-based chemoinformatics strategy to predict relationships between
31  various attributes of ligands and their binding to known targets. These models are increasingly
32  being used in risk assessment and can screen thousands of chemicals for the potential affinity of
33  their 3D structures to bind to active protein binding sites. HTVMD models have been used in the
34  pharmaceutical industry for many years to identify candidate drugs. These models can also be used
35  to estimate the likelihood that a chemical of toxicological interest would bind to a target protein, for
36  example, the potential affinity as a direct agonist of the estrogen receptor.

---

[28]The parentheses around the "Q" in (Q)SAR indicate that the term refers to both qualitative predictive tools, i.e., structure-activity relationships (SARs) and quantitative predictive methods, i.e., quantitative structure-activity relationships (QSARs). Although the term (Q)SAR is often used to refer to predictive models, especially computer-based models, (Q)SAR actually includes a wide variety of computerized and noncomputerized tools and approaches (Hansen et al. 2011).

1  Recent advances in high-performance computing support simultaneous runs of QSAR and HTVMD
2  models, dramatically decreasing the time to discovery. The U.S. Army Medical Research and
3  Materiel Command, for example, has recently published their version of a Docking-based Virtual
4  Screening pipeline that facilitates the usage of the AutoDock molecular docking software on
5  high-performance computing systems (Jiang et al. 2008).

6  The OECD provides a free downloadable QSAR software package, the QSAR Toolbox, that is
7  intended for use by governments, the chemical industry, and other stakeholders to assess potential
8  chemical human and ecological toxicities for data-poor chemicals (OECD 2012). The QSAR Toolbox
9  estimates the potential toxicity of a compound of interest based on the available information (e.g.,
10  mechanism, MOA, or toxicological effects) for structurally similar analogs, and uses read-across or
11  trend analysis to construct categories of chemicals for screening purposes even if only a few of the
12  members in the category have available test data. The popularity of the read-across method is
13  driven by its relative simplicity and the availability of the QSAR Toolbox online. OECD has also
14  developed guidance on the validation of QSAR models when used for regulatory purposes (OECD
15  2004). Assessments informed by new data types and methods will incorporate the results from
16  data sources that can be automated (e.g., QSAR and molecular docking models, and HTS data), with
17  the more traditional data (when available) to advance the speed and accuracy of chemical screening
18  and to support the weight-of-evidence approach to toxicity prediction (Golbraikh et al. 2012, Lock
19  et al. 2012, Rusyn et al. 2012, Wignall et al. 2012, Sedykh et al. 2011). Use of the above models and
20  approaches will advance the ranking of chemicals currently being produced, as well as support the
21  design of new products and chemical processes that increasingly minimize harm to health and the
22  environment.

### 3.3.2. High-Throughput and High-Content Assays

23  HTS and high-content screening (HCS) assays are major tools used for early evaluation of chemicals
24  and their ability to perturb molecular pathways (Judson et al. 2013, Sipes et al. 2013, Tice et al.
25  2013, Kavlock et al. 2012, Judson et al. 2011). Much of the HTS/HCS (for the remainder of this
26  section use of the term HTS includes both HTS and HCS) methodology was developed to aid the
27  pharmaceutical and biotechnology industries in the drug discovery process where one has a drug
28  target of interest (e.g., a receptor or enzyme) and a need to screen up to millions of candidate
29  compounds for leads (Mayr and Bojanic 2009, Bleicher et al. 2003). The technology has been used
30  more broadly in approaches often called chemical genetics (or sometimes chemical biology) where
31  small molecule screening is used to identify probes for biological signaling networks and cellular
32  phenotypes (Schreiber 2003). These assays became of interest in toxicology because many targets
33  of pharmaceutical and chemical biology interest could also be postulated to be involved in disease
34  processes driven by unintentional exposures to environmental chemicals (Houck and Kavlock
35  2008). Generating a large data matrix of toxic chemicals and HTS assays against critical proteins
36  and cellular phenotypic effects provides toxicologists an opportunity to discover novel MOAs that
37  have long eluded the field.

38  The underlying technologies for HTS assays are well known, so a detailed discussion is not
39  presented here. Instead, the discussion focuses on a broad description of the types of assays and
40  some of the key issues to be considered when designing *in vitro* Tier 1 approaches. HTS assays are
41  broadly divided into two types: cell-free/biochemical or cell-based. Cell-free assays typically test
42  for the direct interaction of a test chemical with a specific protein such as a receptor or enzyme.
43  Measures of interaction include binding or inhibition of enzyme activity. In cell-based assays, a

cellular readout can be molecular-based (e.g., changes in gene or protein expression) or phenotypic (cytotoxicity, changes in cell morphology). In a cell-based assay, the selection of the cell system is critical. Assays have been developed using a variety of primary cell types from various organs and species, immortalized cell lines, and stem cell types (Dick et al. 2010). The choices reflect the strengths and weaknesses of the different approaches. For example, immortalized cell lines generally produce very reproducible screening results over long periods of time due to the continuous growth and stability of the cell lines; however, this occurs at the cost of having significant differences from the *in vivo* physiology of the cell type from which the line was derived. The converse holds true for most primary cells, that is, better representation of true physiology but more challenging to work with in producing consistent, reproducible screening results. Co-culture systems combine different cells in an attempt to mimic *in vivo* systems requiring complex cell-cell signaling networks (Berg et al. 2010). Certain whole organisms, including *Caenorhabditis elegans* and zebrafish embryos, can also be used in HTS assays (Smith, MV et al. 2009, Parng et al. 2002).

### 3.3.3. Toxicokinetics

HTS assays provide toxicologists with an efficient and cost-effective tool to broadly screen chemicals for potential proximal biochemical and cellular interactions. As previously mentioned, the HTS assays are run in concentration-response format. The potency of each chemical in each assay can be summarized using $AC_{50}$ or LEC (lowest effective concentration) values, depending on the type of dose-response data collected. The potency values among the *in vitro* assays, along with other chemical information, have been proposed for use in hazard identification (Martin et al. 2011, Sipes et al. 2011) and prioritization of chemicals for further testing (Reif et al. 2010). The relationship between the *in vitro* concentration of the chemical in the well to the concentration of the chemical in the blood or target tissue (*in vivo*), however, can be complex and dependent on variables that are not captured in the HTS assays. These variables include bioavailability, clearance, and protein binding (Wetmore et al. 2012).

*In vitro* to *in vivo* extrapolation (IVIVE) is a process that uses data generated within *in vitro* assays to estimate *in vivo* drug or chemical fate. In the past, IVIVE has been predominantly developed and applied in the pharmaceutical industry to estimate therapeutic blood concentrations for specific candidate drugs, and to identify potential drug-drug interactions (Chen, Y et al. 2012, Shaffer et al. 2012, Gibson and Rostami-Hodjegan 2007). Due to both legislative mandates and public pressure for increased toxicity testing, IVIVE is increasingly being used to predict the *in vivo* PK behavior of environmental and industrial chemicals (Basketter et al. 2012).

A combination of IVIVE and reverse dosimetry can be used to estimate the daily human oral dose (called the oral equivalent dose) necessary to produce steady-state *in vivo* blood concentrations ($C_{SS}$) that are considered equivalent (with respect to chemical concentration at potential targets) to the dose delivered *in vitro* at the $AC_{50}$ or LEC values, and can be used for those values across the more than 600 *in vitro* assays (Wetmore et al. 2012, Rotroff et al. 2010).

### 3.3.4. High-Throughput Exposure Estimation: ExpoCast Prioritizations

The use of HT assays to characterize biological activity *in vitro* enables prioritization of potential environmental hazards once the results of *in vitro* assays have been anchored to, and found to be predictive of, *in vivo* effects. Without capabilities for HT assessment of potential for exposure, prioritization (with respect to potential risk) cannot be completed, as most chemicals have little or no exposure data (Wetmore et al. 2012, Arnot et al. 2010b, Arnot et al. 2010a, Cohen Hubal et al.

1   2010, Rotroff et al. 2010, Hubal 2009, Sheldon and Cohen Hubal 2009, Rosenbaum et al. 2008,
2   Arnot and Mackay 2007, NRC 2006). Currently, few, if any, inexpensive *in vitro* assays are widely
3   available to characterize those properties of chemicals relevant to exposure. Furthermore, the
4   studies for assessing both the presence of environmental chemicals in the immediate vicinity of
5   individuals (exposure potential) and any known biomarkers of actual exposure are expensive, labor
6   intensive, and, with the notable exception of CDC's NHANES, typically difficult to extrapolate to the
7   general population (Rudel et al. 2008, Angerer et al. 2006, Eskenazi et al. 2003). For these reasons,
8   exposure prioritization must be drawn from mathematical models which, when parameterized by
9   chemical-specific properties, provide a structured, consistent way to approach large numbers of
10  unknown chemicals.

11  Physicochemical properties (e.g., water solubility, preference for binding in lipids) inherent to a
12  given compound have been used to predict potential bioaccumulation, and even toxicity, within
13  ecological species to make HT prioritizations of potential chemical exposure (Gangwal et al. 2012,
14  Reuschenbach et al. 2008, Walker et al. 2002, Walker and Carlsen 2002). Beyond inherency,
15  environmental fate and transport models have been developed to account for the accumulation of
16  compounds in various environmental media (i.e., air, soil, water) and the degradation rates of those
17  compounds in those media. These fate and transport models enable predictions of human exposure
18  based on assumptions of human interaction with environmental media and derivation of food from
19  the environment (Arnot et al. 2010b, Arnot et al. 2010a, Rosenbaum et al. 2008, Arnot and Mackay
20  2007). Parameterized using chemical structure and production volumes alone, these models can be
21  used to make HT exposure prioritizations (Arnot and Mackay 2007).

22  EPA is developing the ExpoCast exposure model prioritization framework, which is flexible and
23  expandable to incorporate new HT exposure models as they become available. Currently the
24  framework relies on two quantitative fate and transport models amenable to HT operation: USEtox
25  (Rosenbaum et al. 2008) and RAIDAR (Arnot and Mackay 2007). These models have been
26  empirically assessed for their ability to predict exposures inferred from the NHANES data set.
27  These "ground truth" biomonitoring data are used to calibrate the model predictions and estimate
28  *de facto* uncertainty of the predictions for 41 chemicals where intake per unit emission, total
29  production volume or volume applied, and actual exposures inferred from biomonitoring data were
30  available. The calibration and uncertainty are then extrapolated to ~1,600 chemicals to make rank
31  order predictions on a per unit emission basis, as well as a rank order prediction for ~600
32  chemicals adjusted using production volume (Wambaugh and Shah 2010).

33  NexGen efforts to incorporate exposure prioritization information could proceed along three fronts.
34  First, efforts to evaluate the utility of the predictions must be undertaken to determine if the
35  chemicals of highest priority are indeed present in the environment. Next, new models must be
36  developed to address aspects of exposure currently underrepresented by fate and transport
37  models—namely exposure from personal contact sources (i.e., consumer use). Finally, using the full
38  uncertainty range of the absolute exposure predictions (mg/kg body weight/day), risk potentials
39  could be calculated for risk-based prioritization.

### 3.3.5. Virtual Tissue (VT) Modeling

40  VT models provide an experimental and theoretical framework for the systematic and integrative
41  analysis of complex multicellular systems. These models capture the flow of molecular information
42  across cellular and biological networks, and process this information computationally into higher
43  order responses that ideally simulate a potential adverse outcome(s). Responses to perturbation

| 1 | depend on network topology, system state dynamics, and collective cellular behavior. A unique |
| 2 | aspect is that these simulations are enabled from individual cellular behaviors in a multicellular |
| 3 | field that can result in emergent properties, which are behaviors that arise from interactions of |
| 4 | parts at the next level of a system (e.g., functions, phenotypes) that are not apparent from |
| 5 | knowledge about the behavior of the parts alone. |

| 6 | The field of VT models is in the early stages of development but will become more prominent as the |
| 7 | state of science develops. Jack et al. (2011) and Knudsen et al. (2010) provide examples of VT utility |
| 8 | and the state of science. VT models are practical solutions for translating between biological data |
| 9 | and individual and population-level health outcomes. They combine data and knowledge into |
| 10 | computer models that predict behavior of a complex system, leading to adverse outcomes in |
| 11 | hepatic toxicity, developmental toxicity, reproductive toxicity, cardiopulmonary toxicity, and more |
| 12 | (EPA 2009b). |

| 13 | Virtual models are also briefly discussed in Section 4.4 as one of the new approaches that can |
| 14 | address recurring issues in risk assessment, in this case, dose-response characterization. |

### 3.3.6. Example: Thyroid Pathway Disrupting Chemicals and High-Throughput Systems

| 15 | For EPA to base regulatory decisions on data from mechanistic-based evaluations, several issues |
| 16 | must be addressed. EPA will need to develop criteria and approaches for translating data across the |
| 17 | various types of testing and to identify the types of data and information to support the use of these |
| 18 | data in a regulatory context. To this end, EPA's NexGen Thyroid Disrupting Chemical Workgroup |
| 19 | (EPA 2012a) conducted a thyroid prototype case study that reviewed existing ToxCast™ assays and |
| 20 | provided recommendations for how the data could be used to predict thyroid disruption-induced |
| 21 | developmental neurotoxicity. |

| 22 | A major reason the workgroup selected the thyroid hormone system as its prototype is that the |
| 23 | underlying biology of thyroid hormone homeostasis is well established, thus enabling the |
| 24 | elucidation of the pathway(s) for thyroid hormone disruption (Zoeller and Crofton 2005). The |
| 25 | workgroup identified three issues that should be addressed to use HT assays to predict which |
| 26 | environmental chemicals would likely cause developmental neurotoxicity via disruption of thyroid |
| 27 | hormone homeostasis. These issues are Assay Identification and Refinement; Algorithm |
| 28 | Development for Toxicity and Hazard Prediction; and Standards Development for Assay Conduct, |
| 29 | Data Analysis, and Data Reporting for Risk Assessment Needs. The following is a brief summary of |
| 30 | the case study. |

### Assay Identification and Refinement

| 31 | As a first step, the workgroup identified the HT assays in the ToxCast™ database that assess |
| 32 | endpoints known to be relevant to disruption of thyroid function. The workgroup found that |
| 33 | ToxCast™ contains multiple assays relevant to assessing the potential for a chemical to disrupt |
| 34 | thyroid hormone homeostasis. Coverage of the effects of concern, however, is quite variable. |
| 35 | Although five of the identified assays evaluate endpoints that directly affect the thyroid hormone |
| 36 | pathway (e.g., thyroid hormone receptor binding and TRH receptor binding), the rest evaluate |
| 37 | endpoints not specific to the thyroid hormone pathway. For example, of the 90 assays identified as |
| 38 | thyroid-relevant, 85 are related to hepatic stimulation, metabolism, and clearance of thyroid |
| 39 | hormones. Alteration of these pathways influences thyroid hormone homeostasis indirectly, and |
| 40 | neurodevelopmental effects tied to thyroid disruption by this mechanism are thus secondary effects |

1    of a chemical (inadequate hormone availability due to increased elimination). These secondary
2    effects are in contrast to a primary effect, whereby a chemical interferes directly with the function
3    of the thyroid gland itself or interacts at the site of thyroid hormone receptor in the brain of a
4    developing organism.

5    Adequately assessing the potential of an environmental chemical to disrupt thyroid hormone
6    homeostasis requires that appropriate endpoints be identified and assays be developed and
7    incorporated into testing schemes. This process will involve identifying the specific endpoints in
8    the pathways that need to be tested, additional assays that could be available but which are not
9    currently part of ToxCast™, and additional assays that need to be developed. A recent workshop
10   review by Murk et al. (2013) provides a state-of-the-science assessment of important MIEs for
11   thyroid disruptors, potential and currently used assays for these MIEs, and recommendations for
12   research priorities.

## Algorithm Development for Toxicity and Hazard Prediction

13   The workgroup's second recommendation was to develop algorithms or decision logic flows that
14   balance the potential adversity of the outcome with the uncertainty of the available data. Should
15   assays evaluating endpoints directly affecting the thyroid-related brain changes be weighted more
16   heavily in algorithms than those measuring upstream hepatic enzyme induction? How will
17   algorithms incorporate the fact that multiple chemicals might interact with the same key event, and
18   one chemical might interact with various MIEs, and thus lead to multiple adverse outcomes?
19   Biological plausibility should be the driver in algorithm development.

20   Another aspect to consider is the methods used to incorporate assay results into analyses. Clearly,
21   incorporating many sets of dose-response information into combinatorial analysis requires some
22   simplification of assay results. Many current HT assay results are simplified via classification as
23   either a positive or negative ("hit" or "no hit"), or are assigned a summary statistic such as an $IC_{50}$
24   (the concentration producing a 50% inhibition of response) or lowest effective dose. Obviously,
25   binary decisions such as hit/no hit determinations depend on the criteria chosen to define a hit.
26   These criteria could be derived from statistical significance, biological significance, or an arbitrary,
27   nominal level of change. Depending on the data set, the basis for the classification criteria might be
28   difficult to determine, and might not be consistent across assays. Similarly, summary statistics
29   depend on the model used to generate them or on the specific value chosen (such as $IC_{50}$ versus
30   $IC_{10}$). Relative potency ranks also might vary depending on the shape of the dose-response curve,
31   such that within a given set of chemicals, Chemical A could have the lowest $IC_{50}$ while Chemical B
32   had the lowest $IC_{10}$ value. Lack of such information will lead to greater uncertainty in its use.

## Assay Conduct, Data Analysis, and Data Reporting for Risk Assessment Needs

33   Understanding the characteristics of the individual assays that will serve as the basis of these
34   predictions is critical when using HTS data. Individual assay characteristics are key regardless of
35   the ways in which the data are ultimately used, which might span the spectrum from combinatorial
36   use in predictive algorithms, test batteries for hazard identification and prioritization, to
37   supporting data for individual chemical risk assessments. Although these uses are potentially
38   diverse, several common assay characteristics will be needed. Some of the specific types of
39   information needs might vary depending on the type of risk assessment to be performed.
40   Minimally, the data reporting should include sufficient information to document assay conduct and

1   reliability, the rationale for selection of exposure levels, data analysis techniques, and underlying
2   assumptions regarding assay analysis, conduct, or conclusions.

3   Some advantages of the ToxCast™ data sets are the (1) availability of dose-response information
4   for all assays, (2) availability of assay method details, and (3) availability of the source code for all
5   computational models used in the data analyses. Reliable dose-response information is critical for
6   these types of assays to be useful in risk assessments. Dose-response information is fundamental to
7   understanding the many aspects of chemical toxicity, as it provides a means to evaluate the potency
8   of the chemical and whether a threshold exists.

9   In conclusion, the current case study was complicated by the multitude of target sites at which the
10  thyroid axis can be disrupted (Murk et al. 2013, Crofton and Zoeller 2005); the secondary, indirect
11  nature of the insult produced; and the complexity of the endpoint of concern—neurodevelopment.
12  By conducting this case study, however, the workgroup could identify not only the nodes in the
13  thyroid toxicity pathway that still need coverage, but also the algorithm development and assay
14  conduct issues that should be addressed if HTS assays are to be used in risk assessments.

# 4. Advanced Approaches to Recurring Issues in Risk Assessment

15  In addition to informing chemical specific assessments as discussed above, new data types and
16  advanced approaches also can inform important, recurrent, cross cutting risk assessment issues.
17  These issues are often sources of controversy due to limited data specific to the issue. A number of
18  these issues are discussed below: variability in human response (e.g., genetic variability, early life
19  exposures; exposure to mixtures and nonchemical stressors); inter-species differences; and
20  characterization of low-level chemical exposures likely to be encountered in the environment. This
21  section discussed how new data types and approaches can inform these difficult issues, thus
22  improving our understanding of public health risks.

## 4.1. Human Variability

23  Human response to environmental chemicals is influenced by both intrinsic (e.g., genetics, life
24  stage) and extrinsic (e.g., chemical exposure, stress, nutrition) factors. New methods to examine
25  gene-gene, gene-environment, and epigenome-gene-environment interactions are available (Patel
26  et al. 2013, Lvovs et al. 2012, Meissner 2012, Patel et al. 2012a, Patel et al. 2012b, Baker 2010,
27  Thomas D 2010, Cordell 2009). Zeise et al. (2012) explored how these factors can influence each of
28  the series of biological and physiological steps (known as the source-to-outcome continuum) that
29  ultimately manifests in variability with respect to adverse health outcomes (see Figure 22). The
30  Zeise et al. (2012) review was informed by a National Research Council (NRC) workshop,
31  "Biological Factors that Underlie Individual Susceptibility to Environmental Stressors and Their
32  Implications for Decision Making." The authors considered current and emerging data streams that
33  are providing new types of information and models relevant for assessing interindividual
34  variability.

35  Currently, human variability is usually accounted for by including an uncertainty factor of 1, 3, or
36  10 in the calculation of a reference dose for noncancer health effects. Variability is not explicitly
37  accounted for in cancer health assessment with the exception of the incorporation of an age-specific
38  adjustment factor of ≤ 10 for childhood exposures to genotoxic carcinogens. In a few cases, data on
39  sensitive populations (e.g., asthmatics and those sensitive to air pollutants) might be specifically

1    incorporated into risk assessments. Figure 23 from Zeise et al. (2012) illustrates how different
2    types of variability can influence dose-response relationships.

3    Several strategies have been developed to characterize variability in pharmacokinetics (PKs): (1)
4    for data-rich chemicals (such as pharmaceuticals), a "population PK" approach is used to measure
5    variability and discover the determinants; (2) "predictive PK" uses mechanistic models, assigns
6    *a priori* distributions to specific parameters that can be measured experimentally, and uses Monte
7    Carlo simulations to propagate distributions from model parameters to model predictions; and (3)
8    reproduced "Bayesian PBPK" employs a synthesis of the two previous approaches (EPA 2008).

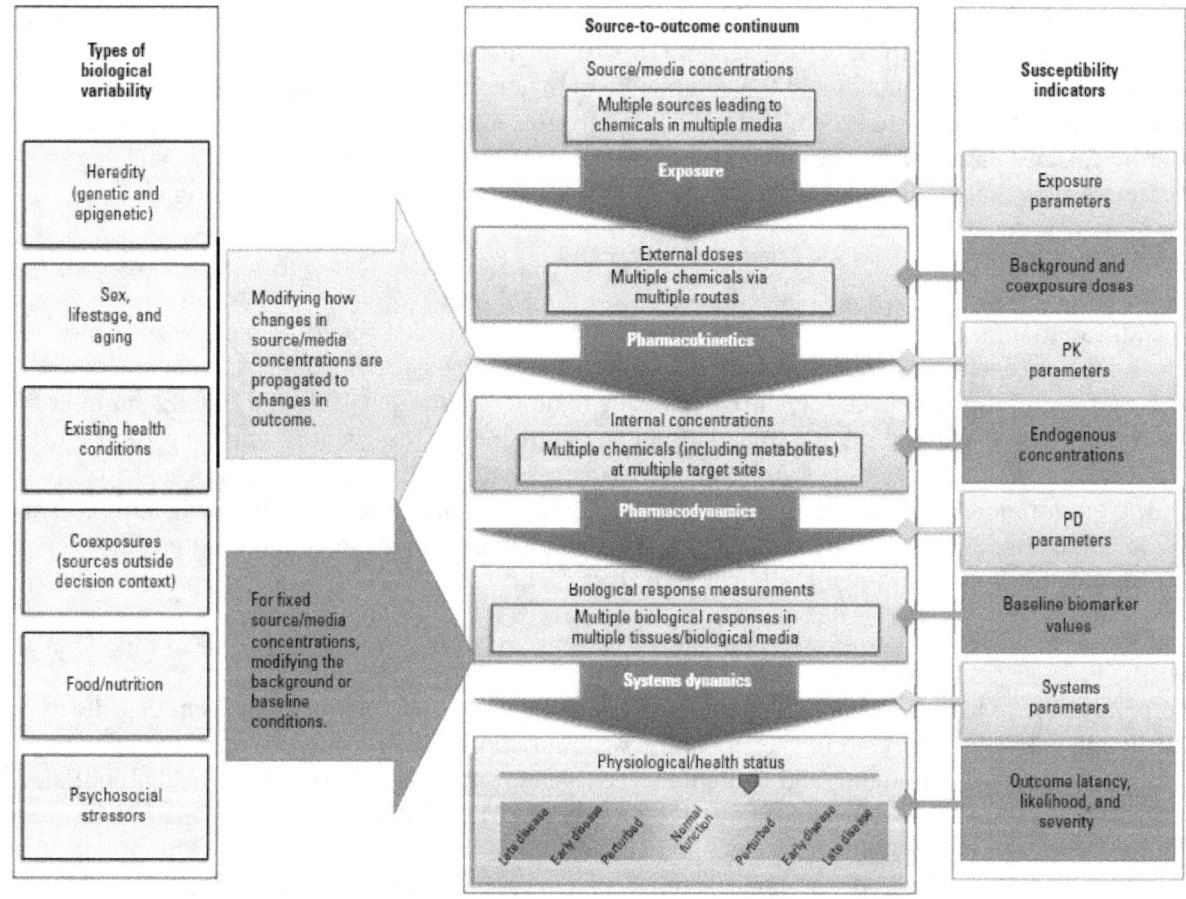

Figure 22. Framework illustration of how susceptibility arises from variability. Multiple types of biological variability intersect with the source-to-outcome continuum, either by modifying how changes to source/media concentrations propagate through to health outcomes, or by modifying the baseline conditions along the continuum. The aggregate result of these modifications is variability in how a risk management decision affects individual health outcomes. The parameters and initial conditions along the source-to-outcome continuum serve as indicators of differential susceptibility, some of which are more or less influential to the overall outcome (see Figure 25) (Zeise et al. 2012). Reproduced with permission from *Environmental Health Perspectives*.

### 4.1.1. Genomic Variability

1    An estimated 20%–50% of phenotypic variation is captured when all single nucleotide
2    polymorphisms (SNPs) are considered simultaneously for several complex diseases and traits. The
3    proportion of total variation explained by individual genome-wide-significant variants has reached
4    10%–20% for a number of diseases (Visscher et al. 2013). Environmental factors are thought to
5    contribute the remaining variability. The interaction between genetic and environmental factors is
6    a key concern in the description of public health risks.

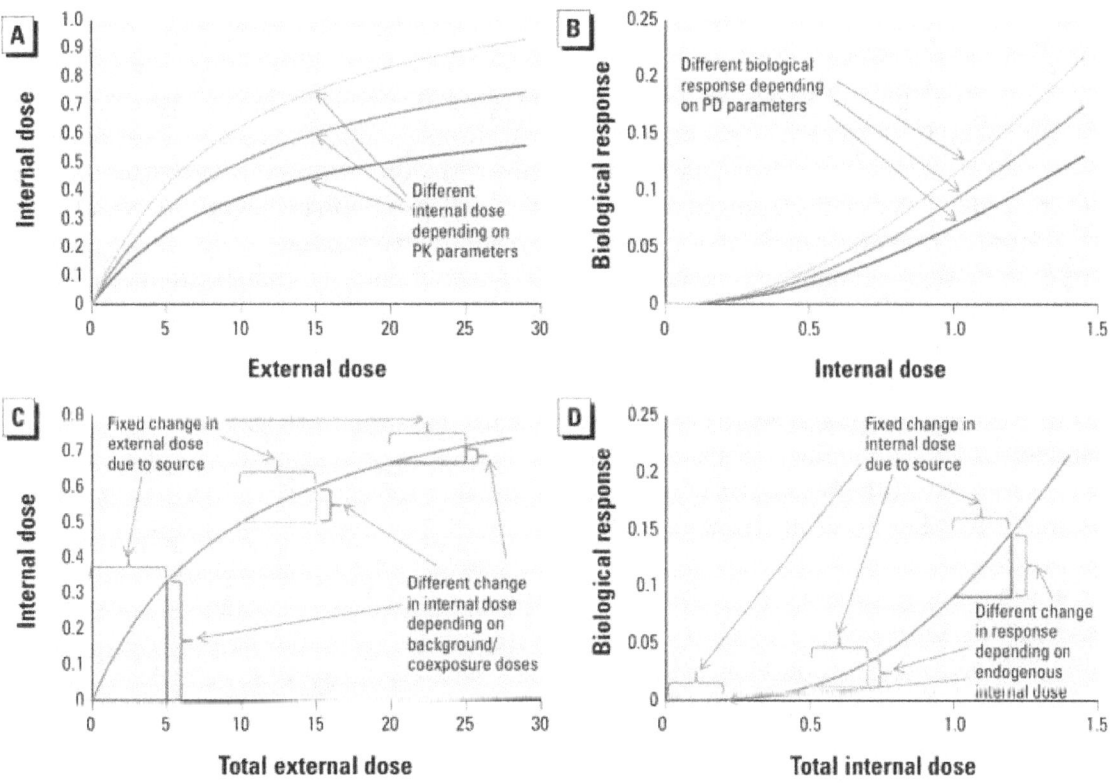

Figure 23. Effects of variability in pharmacokinetics (PK) (A), pharmacodynamics (PD) (B), background/exposures (C), and endogenous concentrations (D). In (A) and (B), individuals differ in PK or PD parameters. In (C) and (D), individuals have different initial baseline conditions (e.g., exposure to sources outside of the risk management decisions context; endogenously produced compounds) (Zeise et al. 2012). Reproduced with permission from *Environmental Health Perspectives*.

7    Several approaches to generating and evaluating genomic data are now emerging that can provide
8    new insights into human variability (both PK and pharmacodynamic [PD]) including (1) *in silico*
9    modeling approaches in which variability in parameter values is simulated, and differences among
10   subpopulations explored (Shah et al. submitted, Knudsen et al. 2011, Knudsen and DeWoskin 2011,
11   Shah and Wambaugh 2010); (2) high-throughput (HT) *in vitro* data generation using cells lines with
12   different genetic backgrounds (Abdo et al. 2012, Lock et al. 2012, O'Shea et al. 2011); (3) *in vivo*
13   studies in genetically diverse strains of rodents to identify genetic determinants of susceptibility
14   (Harrill et al. 2012, NIEHS 2012a); (4) comprehensive scanning of gene coding regions in panels of
15   diverse individuals to examine the relationships between environmental exposures, interindividual

1   sequence variation in human genes, and population disease risks (Mortensen and Euling 2013,
2   NIEHS 2012b); (5) genome-wide association studies (GWAS) to uncover genomic loci that might
3   contribute to human risk of disease (NHGRI 2013, Abecasis et al. 2012, Bush and Moore 2012); and
4   (6) association studies that correlate measures of phenotypic differences among diverse
5   populations with expression patterns for groupings of genes based on co-expression (Friend 2013,
6   Patel et al. 2013, Patel et al. 2012a, Weiss et al. 2012). New understanding of the contribution of
7   epigenomics to disease is rapidly advancing with evaluation of changes such as differential
8   methylation of DNA (Teschendorff and Widschwendter 2012, Hansen et al. 2011, Rakyan et al.
9   2011). Risk assessments of the future will begin to incorporate these types of data as they become
10  available.

11  Panel a) in Figure 24 illustrates one example of how new types of genetic variation data can be used
12  in risk assessment, in this case, how a population concentration-response curve can be estimated
13  for cycloheximide based on HT *in vitro* data using cell lines with different genetic backgrounds. The
14  approach reported by Lock et al. 2012 is being used in Tox21 Phase II, (in collaboration with Rusyn
15  and colleagues at the University of North Carolina) to expand the study of interindividual
16  differential sensitivity to evaluate approximately 1,100 different human lymphoblastoid cell lines,
17  with densely sequenced genomes representing 9 races of humankind, to 180 toxicants. Data will be
18  collected on more chemicals in the future. The numbers of chemicals evaluated in the future in this
19  manner will expand. The large number of human cell lines used allows for an analysis of genetic
20  determinants associated with differential cytotoxicity *in vitro*. This approach will provide
21  significant new insights into human variability in response and can better inform current and
22  future risk assessments. Other examples of human variability data are discussed in the benzene
23  prototype and in Text Box 10 using GWAS data.[29]

---

[29]One caveat: The differential risks conferred by human genetic variability are complex and might not be captured by analyses of small-scale gene variability alone. Hundreds to thousands of genes are likely to be involved in any disease, and multiple variations in genetic makeup might confer similar increased or decreased risk for the same disease. The occurrence of disease also could be influenced by emergent system properties that require analysis not only of how gene variations affect cellular components, but how effects on critical network interactions propagate up through higher levels of the biological system (Torkamani et al. 2008). Consequently, although incorporation of new types of data can better characterize human variability, the characterizations are likely to be incomplete.

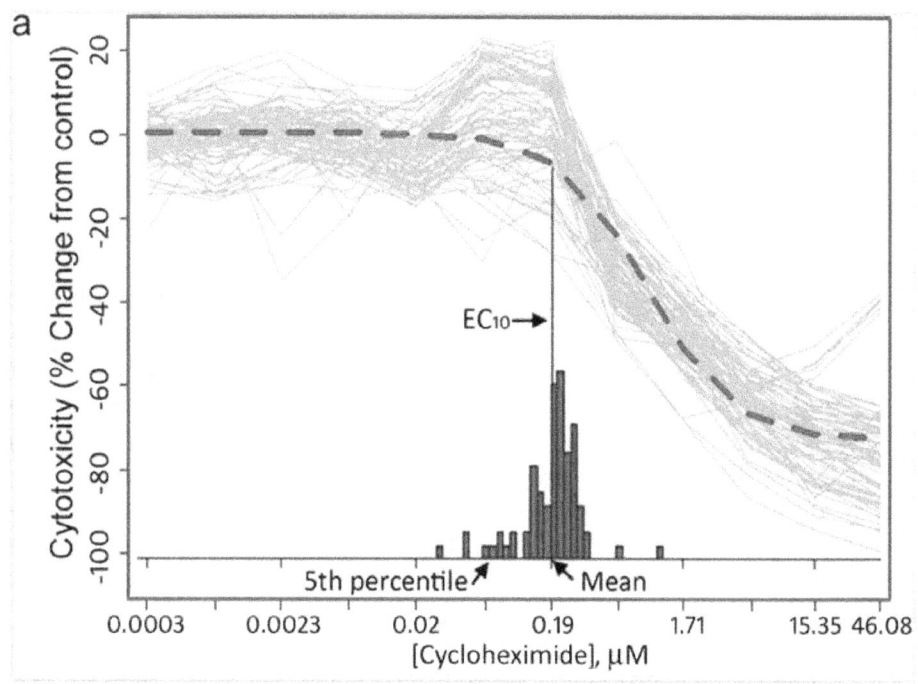

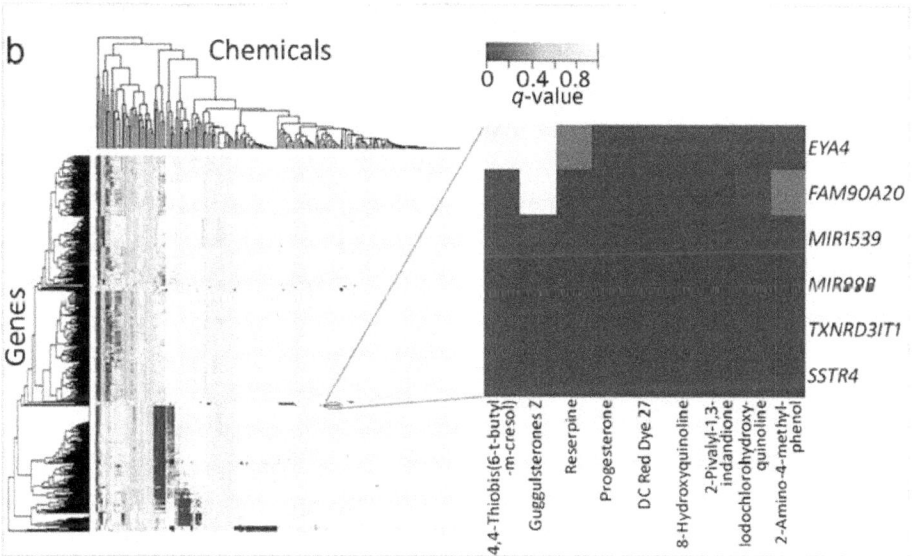

Figure 24. Panel a: A population concentration-response was modeled using *in vitro* quantitative high-throughput screening (qHTS) data using cycloheximide data (cytotoxicity assay) as an example. Logistic dose-response modeling was performed for each individual to the values shown in gray, providing individual 10% effect concentration values ($EC_{10}$). The $EC_{10}$ values obtained by performing the modeling on average assay values for each concentration (see frequency distribution) are shown in the inset. Panel b: A heat map of clustered FDRs (q values, see color bar) for associations of the data from caspase-3/7 assay with publicly available RNA-Seq expression data on a subset of cell lines. A sample subcluster is shown (Lock et al. 2012). Reproduced with permission from Oxford Journals.

### 4.1.2. Early-Life Exposures

1  Early-life chemical exposures can invoke molecular effects that appear to result in increased
2  susceptibility to disease or other morbidity later in life, often via epigenetic modifications
3  (Boekelheide et al. 2012). Evidence from both humans and animals helped establish the influence of
4  early-life exposure on later-life outcomes. For example, human observational data and animal
5  studies report that arsenic exposure during prenatal and early postnatal life increase the risk of
6  cancer, respiratory, and cardiovascular diseases, and neurobehavioral disorders, as supported by
7  human observational data and animal models (Cronican et al. 2013, Boekelheide et al. 2012, Tokar
8  et al. 2012, NRC 2011, Tokar et al. 2011). Later-in-life outcomes can be influenced by time of
9  exposure, species' predisposition to a particular disease, an individual's genetic predilection to
10  disease, or gender. Improved ability to predict disease risk associated with *in utero* or early
11  postnatal exposures results from advances in identifying the targeted genomic region of
12  chemicals/chemical mixtures, epigenetic alteration of gene expression, and the causal links
13  between early-life chemical exposure and later-life outcomes (Boekelheide et al. 2012, NRC 2011).

14  Epigenetic biomarkers for early-life exposures (e.g., placental epigenetic biomarkers, plasma
15  biomarkers) have the potential for use as early indicators of adverse health effects later in life.
16  Development and interpretation of epigenomic biomarkers is in the early stages of development
17  (Hansen et al. 2011, Rakyan et al. 2011); however, as understanding of the underlying epigenetic
18  mechanisms (e.g., DNA methylation, histone modification, microRNA) advances, more will be
19  known about the relationship between biomarkers of early-life exposure and later-life disease risk.
20  A good example is the work that associated early-life exposure to arsenic and DNA
21  hypomethylation with the development of arsenic-induced skin lesions (Boekelheide et al. 2012).
22  The roles of environmental factors that positively and negatively influence health outcomes require
23  study.

### 4.1.3. Mixtures and Nonchemical Stressors

1   Cumulative risk is a function of the exposure to the combined threats from all intrinsic and extrinsic
2   stressors (e.g., chemical exposure, pharmaceutical use, underlying susceptibility, socioeconomic
3   status, work-life stress) and factors that improve health (e.g., good diet, exercise). The assessment
4   of cumulative risk remains a challenging area for human health risk assessment. Only a few studies
5   have examined the potential impact of exposure to environmental chemical mixtures, or to
6   mixtures and nonchemical stressors; and innumerable combinations of chemical mixtures and
7   nonchemical stressors occur in the environment. Conventional methods for risk assessment have
8   made little progress in scaling this particularly mountainous cumulative risk challenge. New
9   methodologies in systems biology, computational models, and data mining provide promise by
10  taking a more comprehensive disease-oriented approach to identification and management of
11  cumulative risk for chemical classes or structures. HTS and omics assay data can be combined with

1 bioinformatics data mining and computational cellular signaling simulations to predict possible
2 disease outcomes (for screening-level assessments) that, combined with higher level systems data,
3 can identify common patterns of significant pathway or network alterations associated with disease
4 (for more quantitative risk assessments). As our molecular understanding of how nonchemical
5 stressors modulate disease continues to evolve, we will also be able to leverage data from systems
6 biology and network analyses to obtain a better understanding of potential cumulative chemical
7 and nonchemical stressor interactions in biological systems and the resulting health impacts.
8 Because epigenomic networks are more easily modulated by environmental factors than the
9 genome, epigenomics should be considered an area of focus for identifying mechanisms that
10 mediate cumulative risks imposed by exposures to environmental factors (Cortessis et al. 2012,
11 Koturbash et al. 2011, Bollati and Baccarelli 2010).

## 4.2. Inter-Species Extrapolation

12 The traditional use of animal models in hazard identification and characterization of dose-response
13 employs chemical testing in mammalian species, and application of an interspecies (animal-to-
14 human) uncertainty factor ($\leq 10$) or body-weight conversion factor to derive an EPA reference
15 value. Increased understanding of the toxicological or biological pathways and their similarity (or
16 lack thereof) among species will improve the extrapolation of chemical effects across species, and
17 the related challenge of selecting model organisms for testing, in contrast to solely comparing apical
18 responses. As knowledge increases on the extent of pathway conservation among species,
19 alternative test species, including nonmammalian vertebrates (adult and embryonic zebrafish) and
20 invertebrate models, will be of greater use in chemical risk assessment. Regulatory toxicology as a
21 whole will move toward increasing reliance on predictive approaches to assessing chemical risk,
22 with a greater emphasis placed on understanding chemical perturbation(s) of conserved biological
23 pathways at key junctures, including molecular initiating events (MIEs) (e.g., activation or
24 inactivation of specific receptors, enzymes, or transport proteins).

25 Data from alternative mammalian species and *in vitro* models are valuable for both ecological and
26 human health risk assessment when used in a pathway-based framework (Ankley, G. T. et al. 2010).
27 The extrapolation between species can occur at different levels of biological organization, such as
28 the MIE, the pathway, and the organ or individual levels. Based on the similarity of pathway-based
29 values to standard toxicological values, this appears to be a useful approach for extrapolating
30 hazard values across species, especially if a known pathway is involved.

31 That gene sequences are conserved—even between distantly related species—is well known and
32 conservation across species is indicative of an essential function. DNA sequence similarity can, but
33 does not always, reflect a functionally conserved role for the genes in question. Investigations of
34 gene function homology can be approached through interspecies comparisons of various
35 components that affect the phenotype in question. The implicated genes, their sequence variation,
36 and the relevant signaling pathways and tissues (cells, organs, circuits) are all informative. Thus,
37 new approaches to understanding the underlying molecular mechanism can improve our cross-
38 species extrapolation (e.g., see Chen et al. (2007), Jubeaux et al. (2012), and Reaume and
39 Sokolowski (2011)).

## 4.3. Low Dose-Response Modeling

1 Empirical dose-response models (e.g., benchmark dose [BMD] models) are widely used in
2 environmental health risk assessment for screening and categorization of toxic substances;
3 determination of toxic potency; determination of a point of departure (POD) for low-dose
4 extrapolation; determination of human exposure guidelines; estimation of risk under specific
5 exposure circumstances; and interpretation of human data. Models that are based on a robust
6 understanding of biological processes, in contrast, are not common. Dose-response models could
7 incorporate data from *in vitro* studies, human or test animal *in vivo* studies, or human epidemiology
8 studies. For public and ecosystem health risk assessment, characterizing population-level
9 responses is the goal.

10 Many risk assessments require models that can extrapolate beyond the data set used in developing
11 the model to derive the toxicity values of interest. Such models are called biologically based models.
12 To date, the main biologically based models used in risk assessment are physiologically based
13 toxicokinetic (PBTK) models that simulate the toxicokinetic behavior of a chemical (i.e., the internal
14 disposition of the chemical in the body following a given dosing regimen). Only a few examples of
15 physiologically based toxicodynamic (PBTD) models are available to characterize the "response"
16 side of the dose-response curve. Well-developed and adequately tested PBTK models are currently
17 used in risk assessment to simulate the toxicokinetics of a chemical or chemicals across dosing
18 regimens (duration, amounts, delivery rate, routes) and species, or from *in vitro* regimens to *in vivo*
19 doses (IVIVE).

20 The establishment of human exposure guidelines for environmental agents involves determining a
21 POD on the dose-response curve, such as a particular response level on a BMD model estimate of
22 the dose-response, corresponding to a specified increase in risk usually in the 5% to 10% range, or
23 signal-to-noise-crossover dose introduced by Sand et al. (2011). This POD is then further reduced
24 by adjustment factors to derive a level of exposure that is considered to be protective of human
25 health and the environment. The National Research Council (2009) suggests an integrated
26 approach to the establishment of human exposure guidelines using adjustment factors applied to
27 the POD, where the magnitude of the factor depends on the "expected" behavior of the exposure-
28 response curve at low levels of exposure. The NRC also examined the influence of background
29 exposures and background disease rates on the shape of the exposure-response curve at low levels
30 of exposure.

31 Characterizing the expected response at low exposure levels (i.e., those that the public is most likely
32 to encounter) is another of the great challenges to previous methods used in risk assessment,
33 specifically the use of relatively high-dose *in vivo* animal assays as the source of data for apical
34 endpoints because the spectrum of adverse effects might be quite different at lower doses. The NRC
35 (2007) recommended developing new approaches and models to generate the data needed for
36 characterizing dose-response curves and to improve estimates especially at doses applicable to
37 likely human exposures. Examples of some new approaches to dose-response modeling are
38 described in Burgoon and Zacharewski (2008), Parham et al. (2009), and Zhang et al. (2010). The
39 application of sensitive HTS assays for pathway perturbations that directly measure biological
40 effects at environmental exposure levels are described in Rotroff et al. (2010) and Wetmore et al.
41 (2012). The reduced cost of HTS assays relative to mammalian toxicity tests might also permit the
42 use of a much broader range of exposure levels, leading to a more detailed description of dose-

1     response relationships throughout the exposure range of interest. Figure 25 summarizes the
2     automated dose-response modeling approach proposed by Burgoon and Zacharewski (2008).

3     A new class of biologically based models called "virtual models" is also being developed to simulate
4     normal biology and to predict how chemical perturbations might lead to adverse effects (i.e., to
5     predict a chemical's toxicodynamics) based on knowledge of potential mechanisms. Examples of
6     virtual models being developed at various levels of biological organization or function include
7     (1) the Physiome Project (Physiome Project 2013), a major resource and model repository of
8     hundreds of physiology models (Hunter et al. 2002); (2) the European Virtual Physiological Human
9     (VPH) project (Hunter et al. 2010); (3) HumMod, a whole-body integrated human physiology model
10     (Hester et al. 2011); (4) Virtual Cell (V-Cell), a spatially realistic quantitative model of intracellular
11     dynamics (Moraru et al. 2008); (5) EPA's Virtual Embryo™ (v-Embryo) project, a suite of models
12     that simulate normal development leading to the formation of blood vessels, limb-buds,
13     reproductive systems, and eye and neural differentiation (Knudsen et al. 2011, Knudsen and
14     DeWoskin 2011); (6) EPA's Virtual Liver™ (v-Liver) model that simulates the dynamic interactions
15     in the liver used to translate *in vitro* endpoints into predictions of low-dose chronic *in vivo* effects in
16     humans (Shah and Wambaugh 2010); and (7) the Virtual Liver Network (German Federal Ministry
17     for Education and Research 2013), a German initiative to develop a dynamic model of human liver
18     physiology, morphology, and function integrating quantitative data from all levels of organization
19     (Holzhutter et al. 2012).

1 The results from large and essential areas of research, including epigenomics, are rapidly adding to
2 our knowledge and will be incorporated in more detail in future efforts. [30]

3 The three sets or tiers of prototypes were intended to explore different aspects of decision context.
4 The primary intent of the first set of chemicals (Tier 3 prototypes) was to verify if and how new
5 data and approaches could be used to inform risk assessment by comparison to robust traditional
6 assessments where risks are generally considered "known." In essence, we attempted to reverse
7 engineer from known answers to verify new approaches, explore value information, and begin to
8 characterize decision rules that could be reasonably applied to chemicals with limited or no
9 traditional data. Secondarily, the Tier 3 prototypes explored how new types of data could expand
10 our understanding of well-studied chemicals. The intent of the Tier 2 prototypes was to explore
11 new types of computational analyses and short-duration *in vivo* bioassays that are intermediate in
12 terms of required resources and confidence in the data between Tiers 3 and 1, and are suitable for
13 evaluating hundreds to thousands of chemicals. These approaches are relatively uncommon in risk
14 assessment to date but hold much promise. The intent of the Tier 1 prototypes was to explore
15 entirely HT approaches that could be applied to tens of thousands of chemicals, might have the
16 greatest uncertainties, but are the least resource intensive to use.

17 The following eight chemicals or chemical classes and their associated effects were chosen for
18 prototype development:

19 • Tier 3:

20   o Benzene and leukemia (molecular epidemiology),

21   o Ozone and lung inflammation and injury (molecular clinical studies), and

22   o Benzo[a]pyrene (B[a]P, a polycyclic aromatic hydrocarbon (PAH) and liver cancer
23     (molecular clinical studies meta-analyses and *in vivo* rodent bioassay).

---

[30]In terms of top-down approaches, molecular, computational, and systems biology data have
grown phenomenally in recent years, and have informed mechanisms of disease and factors that
alter risks of disease. These data are generally stored in large databases such as ENCODE, Gene
Expression Omnibus (GEO), and the Comparative Toxicogenomic Database (CTD) and are publicly
available for further analyses. Analyses and meta-analyses of these data are providing new insights
into environmental public health risks. Bioinformatics (computer-assisted approaches) are
necessary to use these new data effectively due to the size of the relevant databases. The polycyclic
aromatic hydrocarbon (PAH) and diabetes prototypes, in particular, illustrate bioinformatic
"knowledge mining" to understand environmentally related disease.

In terms of bottom-up approaches, many new high- and medium-throughput methods have been
and are being developed that facilitate testing and evaluation of chemicals on an unprecedented
scale. In particular, the *in vitro* evaluations of chemicals with limited or no traditional data are being
enabled. ToxCast™ and Toxicology in the 21st Century (Tox21) provide examples (see Section 3.3).
Tox21 will test ~10,000 chemicals in a few years. New *in vivo* short-duration (hours to weeks)
exposure paradigms also are emerging that provide new types of data to be used in health
assessments. These paradigms use both nonmammalian (see Section 3.2.2) and mammalian species
(see Section 3.2.3).

1   • Tier 2:

2       ○ Chemicals associated with diabetes and obesity ("big data" knowledge mining),

3       ○ Chemicals associated with thyroid hormone disruption (short-duration *in vivo*
4           exposure bioassays – alternative species), and

5       ○ Chemicals associated with cancer (short-duration *in vivo* exposure bioassays –
6           mammalian).

7   • Tier 1:

8       ○ Chemicals associated with cancer and noncancer disorders, especially
9           developmental (QSAR) and

10      ○ Chemicals associated with thyroid hormone disruption (high-throughput *in vitro*
11          assays).

12  Table 9 provides more information on methods explored in each prototype (Krewski et al. 2013).

## 5.2.  5.2.    Implications for Risk Assessment Derived from Prototypes

13  Based on the prototypes provided here and the work of others, new molecular, computational, and
14  systems biology tools likely can better inform risk assessment. Substantial caution in interpretation
15  and use of new information is warranted, however, in large part because our understanding of the
16  science is still evolving, and appropriate data are still scarce. We propose initially to use new
17  methods discussed in this document to: (1) generate hypotheses; (2) screen and rank chemicals for
18  additional research and assessment; and (3) augment understanding of traditional data. Areas of
19  particular promise include improved understanding of relative potency of chemicals to disrupt
20  biologic processes, hazard identification, and mechanisms of disease and disorders; human
21  variability and susceptibility; human relevancy of animal models; and low-dose-response
22  relationships. These future risk assessments ideally would rely on the integration of a variety of
23  new types of data and traditional data, as available. Additional discussion of the lessons learned
24  from the prototypes follows.

25  Systems biology context is key to understanding these new data types and the relationship among
26  various types of data. Network-level understanding is typically more informative than pathway-
27  level understanding, which is usually more informative than individual genes. In general,
28  information on individual genes, in the absence of systems biology-level of understanding, is likely
29  to be inadequate for risk assessment purposes. Information that links molecular events to apical
30  outcomes need not be chemical specific, but can be derived from mechanistic information on
31  disease or from related chemicals. As with any risk assessment, the studies used should be well
32  designed, conducted, and reported; systematic review criteria are necessary in study selection.
33  Characterization of multisource variability is a substantial challenge with new data types because of
34  the sheer amount of data being analyzed and, thus, must be carefully considered. Also, traditional
35  weight-of-evidence criteria continue to be useful in considering new data types, for example, data
36  from multiple, similar studies are preferred (Krauth et al. 2013). That environment-induced
37  changes in biology are dynamic in nature also should be noted, and these dynamic changes are not
38  well understood.

1    Highlights of the prototypes include:

2    •   The effects of human chemical exposures at environmental levels on molecular events were
3        linked to intermediate biological events and apical adverse outcomes using molecular
4        epidemiology, molecular clinical, and environment-wide association studies (e.g., evaluation
5        of NHANES) (EPA 2013, Patel et al. 2013, Devlin 2012, McHale et al. 2012, Patel et al. 2012a,
6        Thayer et al. 2012, Burgoon 2011, McHale et al. 2011, Smith, MT et al. 2011). Chemicals
7        evaluated included benzene, ozone, B[a]P (a PAH), metals, and persistent organic pollutants.

8    •   The B[a]P and diabetes prototypes illustrated the use of "big data" knowledge mining to
9        identify associations between environmental chemical exposures and disease (Patel et al.
10       2013, Patel et al. 2012a, Burgoon 2011). The chemicals of concern for diabetes identified
11       using knowledge mining also were identified in a review of traditional literature by experts
12       (Thayer et al. 2012). This powerful, relatively new technique has not been used extensively in
13       environmental risk assessment, although it is commonly used in other areas of biology.
14       Knowledge mining is particularly useful in developing a broad understanding of potential
15       mechanisms of action, factors that may cause or modify disease risks, and human variability
16       and susceptibility.

17   •   Short-duration exposures coupled with new molecular and computational approaches appear
18       to provide additional insights into potential environmental risks. Use of both alternative
19       species and mammalian species in these new experimental models is explored. These models
20       are faster and less expensive than the molecular epidemiology and molecular clinical studies
21       noted above. Furthermore, unlike the QSAR and high-throughput (HT) models noted below,
22       these models address intact metabolism and cell, and tissue interactions and can be used to
23       study more complex outcomes such as developmental and neurobehavioral outcomes. In the
24       case of alternative species, these models can detect effects over the entire lifespan of the
25       organism and to population dynamics. These models have been used successfully to describe
26       mechanisms, explore complex mechanistic behaviors, describe hazards, and evaluate
27       chemical potency. Confidence in the data also generally lies between Tier 3 and Tier 2
28       approaches (Perkins et al. (2013), Thomas RS et al. (2013a), and Padilla et al. (2012).

**Table 9. Prototype Use of New Scientific Tools and Techniques (Krewski et al. 2013)**

| Scientific Tools Used in Specific Prototypes | Tier 1: Screening and Prioritization | | Tier 2: Limited Scope Assessments | | Tier 3: Major Scope Assessments | |
|---|---|---|---|---|---|---|
| | Cancer & Hydrocarbon Mixtures (QSAR) | Endocrine Disruptors & Deep Water Horizon Oil Spill Dispersants (In Vitro Bioassays) | Diabetes & Multiple Stressors (Knowledge Mining) | Cancer & Reproductive/ Developmental Hazards (Short-Duration In Vivo Exposure Bioassays) | Lung Injury & Ozone (Molecular Clinical) | Leukemia & Benzene (Molecular Epidemiology) |
| **Hazard Identification and Dose-Response Estimation Methods** | | | | | | |
| Quantitative structure-activity relationships | ■ | ■ | | ■ | | |
| High-throughput *in vitro* assays | | ■ | ■ | ■ | ■ | ■ |
| High-content omic assays | | | | ■ | ■ | ■ |
| Molecular and genetic population-based studies | | | | | ■ | ■ |
| Biomarkers of effect | | | ■ | | ■ | ■ |
| Pathway/network analyses | ■ | ■ | ■ | ■ | ■ | ■ |
| **Dosimetry and Exposure Assessment Methods** | | | | | | |
| *In-vitro-to-in-vivo* extrapolation | ■ | ■ | | ■ | | |
| Pharmacokinetic models and dosimetry | | ■ | | ■ | ■ | ■ |
| Biomarkers of exposure | | | ■ | | | |
| **Cross-cutting Disciplines** | | | | | | |
| Bioinformatics/ computational biology | ■ | ■ | ■ | ■ | ■ | ■ |
| Functional genomics | | | ■ | ■ | ■ | ■ |
| Systems biology | | | ■ | ■ | ■ | ■ |

1    • QSAR models (Venkatapathy and Wang 2013, Goldsmith et al. 2012, 2012a, Wang, N et al.
2      2012b) and HT *in vitro* bioassays are being used to rapidly evaluate a wide array of chemicals
3      (Judson et al. 2013, 2011, Sipes et al. 2013, Tice et al. 2013, Kavlock et al. 2012, Rusyn et al.
4      2012). "These tools can probe chemical–biological interactions at fundamental levels,
5      focusing on the molecular and cellular pathways that are targets of chemical disruption"
6      (Kavlock et al. 2012). Thousands of chemicals are currently being evaluated, particularly in
7      the ToxCast and Tox21 programs. Both estimates of potency and insights into potential
8      hazards are being generated. Additionally, tools exist to relate *in vitro* concentration to
9      potential human exposure levels (reverse dosimetry) (Wetmore et al. 2013, Wetmore et al.
10     2012, Rotroff et al. 2010, Hubal 2009). Although directly correlating *in vitro* findings to risks
11     of human disease is difficult, these QSAR and HT methods provide powerful new tools for
12     screening and ranking large numbers of chemicals for further evaluation and assessment, as
13     well as exploring underlying mechanisms of toxicity, and evaluating human variability in
14     response to chemical exposures (Lock et al. 2012).

15   Thomas RS et al. (2013a) propose a framework for incorporating these new technologies into
16   toxicity testing and risk assessment in an integrated fashion. The first steps proposed are to use
17   *in vitro* assays to separate chemicals based on their relative selectivity in interacting with biological
18   targets and to identify the concentration at which these interactions occur. Dosimetry modeling
19   converts *in vitro* concentrations into external dose for calculation of the point-of-departure (POD)
20   and comparisons to human exposure estimates to yield a margin of exposure (MOE). The second
21   step involves short-term *in vivo* studies, expanded pharmacokinetic evaluations, and refined human
22   exposure estimates, thus increasing confidence in the evaluation. The third step represents the
23   traditional animal studies currently used to assess chemical risks. A significant percentage of
24   chemicals evaluated in the first two tiers could be eliminated from further testing based on their
25   MOE. Additionally, at each step, information might be suitable for supporting some types of Agency
26   decision-making. The framework provides a risk-based and animal-sparing approach for evaluating
27   chemicals using technological advances to increase efficiency.

28   In addition to informing hazard identification and dose-response, new data types and methods have
29   the potential to inform recurrent, challenging risk assessment issues.

30   **Experimental Low Dose Data vs. Low Dose Extrapolation** – Dose-dependent molecular changes
31   associated with adverse outcomes can be observed at environmental concentrations. Thus, these
32   new approaches can provide experimental data to help characterize dose-response relationships at
33   concentrations where responses, to date, have often been only inferred. Both assay methods and
34   statistical analyses must demonstrate sufficient sensitivity to be considered informative. Observed
35   molecular changes include changes in both magnitude and character, reflecting underlying
36   alterations in biology with increasing dose and time. Biological processes that are consistently
37   observed across the exposure range of interest are likely to be the most useful as biomarkers of
38   exposure and effect. Elucidating the meaning of these dynamic changes in terms of risk will be
39   challenging.

40   **Variability and Susceptibility** – New data and methods can enhance our ability to understand
41   variability in response and the identification of potentially susceptible populations. Human cells
42   from various individuals (e.g., 1000 Genome Project) evaluated in *in vitro* high-throughput models
43   provide an avenue for understanding responses across subsets of the human population (Lock et al.
44   2012). Data mining and bioinformatics analyses will facilitate the identification of susceptible

1  populations and underlying sources of variability by combing existing molecular epidemiology and
2  clinical databases. In all, this work can provide quantitative data, which to date have been generally
3  lacking, to support more accurate estimates of human variability and identification of susceptible
4  populations.

5  **Evaluation of the Effects of Multiple Stressors –** The ability to map mechanism of disease and
6  adverse outcome pathways disrupted by various environmental agents gives us new tools for
7  understanding the interactions of multiple environmental stressors, including chemical mixtures
8  and lifestyle factors.

9  **Certain caveats** that apply generally to use of new data types in risk assessment deserve mention.

10 • Cell type, tissue, individual, subpopulation, species, and test system can alter how specific
11   omics are expressed as traditional intermediate and apical outcomes, even when the
12   molecular signature is the same. This is likely due, at least in part, to epigenomic differences
13   and genomic plasticity. This issue should be considered, as feasible, in data interpretation.

14 • The metabolism of many chemicals often plays an important role in toxicity. That most HT
15   test systems are not metabolically competent is important to consider. Various approaches to
16   the issue of *in vitro* metabolism are being evaluated; however, this currently remains a
17   complicating factor in most *in vitro* testing.

18 • Molecular profiles appear time-dependent, that is, they evolve over time with continued
19   exposure and post-exposure. This can confound prediction of outcomes or disease outcomes
20   based on "snapshots" in time of biological events. Fortunately, however, at least some
21   signatures appear to stabilize over time and can serve as reliable indicators of chronic
22   outcomes.

23 • Currently, studying multiple molecular processes (i.e., genomics, transcriptomics, proteomics,
24   and epigenomics) in a single study is relatively rare, primarily due to expense. This lack of
25   biological integration limits our understanding.

26 • Due primarily to experimental design and reporting issues (see B[a]P [a PAH] prototype),
27   adequate data from the open literature to support risk assessment activities currently are
28   available for few chemicals. This underscores the importance of high-quality research and
29   testing programs like ToxCast™ and Tox21 and systematic review of data.

30 • Data reproducibility and false negative rates may remain a potential limitation of high
31   throughput screening and high content assays (e.g., toxicogenomics). The false negative rate
32   (i.e., calling a chemical non-toxic when it is) tends to decrease as an increasing number of
33   independent replicates are used. Successful screening programs need low false negative rates,
34   while balancing their efficiencies (i.e., cost, time, throughput).

35 The challenge is to use what we know today wisely, with the understanding that biological
36 knowledge is evolving very rapidly, and likewise, risk assessment also will need to evolve.

## 5.3. Summary

37 Throughout this report, examples are provided that illustrate how new types of data might be used
38 to improve risk assessment. Table 10 summarizes (1) various decision context examples common
39 at EPA; (2) a "toolbox" of various NexGen methodologies that could provide data to support each
40 decision context; (3) types of "fit for purpose" toxicity values that might be derived from new data

1     types or traditional data; and (4) assessment products in which molecularly or computationally
2     informed toxicity values could be used. Although all approaches can be used in any type of
3     assessment, any one of the health data approaches listed in Tiers 1 and 2 could provide a minimum
4     data set. In this scheme, Tier 1 is primarily QSAR or HT data-driven. Tier 2 is high-content or
5     traditional data-driven (in addition to Tier 1 data, if available). Tier 3 will continue to be traditional
6     data driven but could be augmented by molecular, computational, and systems biology data if the
7     data are available, of sufficient quality, and substantively useful.

**Table 10. Problem Formulation Table**

| PROBLEM FORMULATION | | |
|---|---|---|
| **Tier1: Screening and Prioritization** | **Tier 2: Limited Scope Assessments** | **Tier 3: Major Scope Assessments** |
| **Decision Context Examples**<br>• Emergency response<br>• Unregulated drinking water chemicals identification<br>• Potential emerging chemical problems or opportunities<br>• Research directions | • National Air Toxics Assessment<br>• Superfund listing/removal actions<br>• Drinking Water Health Advisories | • National Regulatory Decisions<br>• International, Tribal, State, & Local Technical Support |
| **Toolbox of Possible Approaches**<br>• QSAR<br>• High-throughput (HT) Screening Assays<br>• Computational Toxicology Models<br>• No Traditional Data<br>• Automated Data Integration | • High-content Assays<br>  ➤ Knowledge Mining<br>  ➤ Short Duration *In Vivo* Exposure Paradigms[a]<br>• Limited Traditional Data[b]<br>• Automated Data Integration | • Molecular Biology Data<br>• Systems Biology Data<br>• All Policy Relevant Data<br>• Hand-Curated Data Integration |
| **Possible Types of Toxicity Values**<br>High-Throughput Toxicity Values | High-content Toxicity Values | Molecularly Informed Traditional Type Values |
| **Health Assessment Categories**<br>Prioritized Chemicals of Concern List; Screening Values | Provisional Toxicity Values | IRIS or ISA |
| **Exposure Assessment**<br>Physical-Chemical Surrogates | Limited Exposure Data | Extensive Exposure Data |

Increasing Evidence →

[a] Both alternative and mammalian species paradigms.
[b] Potentially not chemical-specific data but rather disease or chemical class data.

8     Integration of information from multiple data types is preferred, but all types of data shown for any
9     tier might not be available or of sufficient quality for inclusion in an assessment. Systematic review
10    criteria are being established and are discussed in section 3.1.3 (McConnell and Bell 2013).
11    Stakeholder input and external peer review will be solicited for new approaches to risk assessment.

12    Systems biology understanding is a fundamental aspect of the weight-of-evidence evaluation. As
13    one progresses from Tier 1 to Tier 3 assessments, the weight of evidence increases; however, the

1  resources to generate the assessments also increases. For example, in Tier 1, toxicity values can be
2  generated solely from extant QSAR data, a process that can be fully automated to be very quick and
3  cost-efficient for a large number of chemicals. Wignall et al. (2013)(SOT poster abstract; manuscript
4  in progress) describe an approach to generate toxicity values for chemicals with limited
5  experimental data using a combination of QSAR, regression, and hybrid modeling (Rusyn et al.
6  2012), and incorporating Organization of Economic Co-operation and Development (OECD)
7  principles for model building and external cross-validation. Tier 2 type assessments, ideally, enable
8  the use of more types of data to inform our understanding of data-limited chemicals. For Tier 2, EPA
9  is beginning to develop high-content toxicity values on a trial basis. High-content toxicity values can
10 be developed using bioinformatic approaches based on data that can be machine read and, hence,
11 readily mined and analyzed. Additional data integration and hand curation might be needed to use
12 available data resources, such as high-throughput screening (HTS)/high-content screening (HCS)
13 assays, alternative species testing, and study data compiled the European Union's Registration,
14 Evaluation, Authorisation and Restriction of Chemicals (REACH) regulation. Tier 3 assessments will
15 continue to be driven by traditional data, but new data types could provide new insights into
16 difficult issues such as low dose-response, human variability and susceptibility, and the effects of
17 multiple environmental stressors. These various "fit for purpose" assessment types can be used to
18 develop hypotheses, screen chemicals, mechanistically fingerprint toxicants, set priorities, and
19 inform hazards, relative potencies, and risks.

## 6. Conclusions

### 6.1. Challenges

20 Novel data streams and approaches are rapidly emerging that present opportunities for informing
21 and supporting human health risk assessment, but challenges remain. Four key challenges are the
22 need for (1) the ability to predict metabolism of test compounds, (2) improved understanding of
23 the biology from a systems perspective; (3) evaluated methods to measure key aspects of biological
24 space across multiple levels of organization; and (4) the knowledge infrastructure to ensure
25 availability of relevant data. Future directions include filling these scientific gaps and continuing to
26 build the framework for incorporating new information fit-for-purpose into assessments to support
27 a range of decisions to promote health, protect the environment, and manage risks.

28 Arguably, the greatest challenge is posed by the need to consider and evaluate complex interactions
29 of chemical and biological systems to predict potential for health risks. Systems biology provides an
30 approach for investigating emergent properties in complex chemical-biological systems by probing
31 how changes in one part affect the others, and the behavior of the whole. New data types are
32 providing required information to develop these predictive models.

33 There is an imbalance, however, in the sophistication of methods available and the resolution of
34 data being developed to evaluate impacts of chemical perturbations and to discover mechanistic
35 commonalities. Large amounts of network or high-throughput screening/high-content data can be
36 collected to measure effects at the molecular level. Substantial information is also available on
37 disease outcomes, yet only very sparse data are being generated on intermediate events. A similar
38 lack of exposure information commensurate with hazard data is also evident. Even given the rich
39 data coming from implementation of the high-throughput (HT) toxicity testing schemes, gaps in
40 coverage for key endpoints occur, and thus, developing and incorporating assays are needed to fill
41 gaps in the biology required to assess potential for the full range of adverse outcomes required by

1 risk assessors. This discrepancy in available data across levels of biological organization should
2 narrow over time, as methods continue to advance and as more metabolomics data for biomarkers
3 of effects and exposure are made available. This will lead to development of models that predict
4 disease outcomes with greater certainty from initiating events in individuals and populations
5 relative to exposures likely to be experienced in the real environment.

## 6.2. Next Steps

6 Future plans for facilitating use of new data types and tools to support the full range of risk-based
7 assessments and decisions include addressing needs for validated testing schemes and clearly
8 articulating decision considerations for incorporating results of these analyses. In addition, further
9 prototypes or case examples for incorporating HT toxicity data and other novel data types to inform
10 risk assessment are required to demonstrate the added value of these advanced tools and to
11 identify further the most significant scientific gaps.

12 Validation of HT toxicity testing schemes will be necessary if the data developed using these
13 methods are to be used to inform risk-based decisions and to support efficient chemical risk
14 assessments. The key to moving the wealth of information being generated through research efforts
15 such as ToxCast™ and Toxicology in the 21st Century (Tox21) is to develop a framework for
16 validating HT toxicity testing schemes to support specific chemical evaluation objectives.
17 Traditional "validation" schemes designed to evaluate conventional assay and testing structures do
18 not adequately address this gap and would take years to implement. As the technology for
19 providing rapid, efficient, robust hazard and effects data continues to advance, the validation
20 process for evaluating these new methods is also expected to undergo a transformation to provide
21 fit-for-purpose confidence in results. Future incorporation of new types information to improve the
22 scientific basis and efficiency of risk assessment requires clear articulation of decision
23 considerations for using new types of data and methods. Some of these decision considerations
24 might have standard principles supported by a broad range of risk managers and stakeholders,
25 while others will need to be fit-for-purpose. Early consideration of these decision considerations
26 has been initiated and plans are in place to develop criteria for systematic review of new types of
27 data, disease signatures, adequate weight of evidence for use in risk assessment, and new
28 approaches for risk assessment.

29 Demonstrating approaches for incorporating new molecular biology data and evaluating advanced
30 methods might be facilitated by additional case examples and prototypes. Conducting a variety of
31 case studies focused on using the HT toxicity data from ToxCast™ and Tox21, in combination with
32 other chemical-specific information to improve efficiency of risk-based decisions where little
33 traditional toxicity data are available, will be important for assessing the value added of these new
34 data types.

35 Examples also will be identified where molecular biology data can be considered for Tier 3
36 assessments to augment traditional assessment methodologies. These will provide opportunities to
37 solicit public comment and peer review.

38 Opportunities also exist for using new data types to guide development of NexGen approaches by
39 considering prototypes for how this information could support some of the most challenging
40 questions faced by risk managers. Population-level risks could be considered using both traditional
41 and molecular biology data, with an additional emphasis on epigenomics and influences of broadly
42 defined environmental factors. Additional insights for risk managers can be found in Crawford-

1    Brown (2013). Application of new methods might better inform our understanding of the combined
2    effects of multiple stressors, such as multiple chemical exposures, diet, stress, and pre-existing
3    disease. In recognition of the tremendous potential for these new methods and data types to
4    support risk assessment, the EPA Office of Research and Development will continue to elaborate the
5    NexGen framework, and begin to develop toxicity values informed by new biology for specific risk
6    assessment purposes.

7    • EPA's Office of Research and Development will work with EPA's Program Offices using Tier 1
8      screening and prioritization approaches to queue up new assessments. Results from this work
9      will be used to feed back into the testing paradigm for its refinement.

10   • Toxicity values informed by new types of knowledge will be developed in each tier to address
11     needs from screening chemicals for future testing to assessment for potency or category of
12     adverse effect.

13   • Levels of confidence in those values will be characterized depending on the types and quality
14     of the supporting data.

15   EPA's Office of Research and Development will expand stakeholder discussion and the community
16   of practice with regard to the use of new data types and methods in risk assessment, and the peer
17   review of new methods. New assessments will receive public comment and peer review.

18   Finally, EPA's Office of Research and Development will continue working with other national and
19   international agencies involved in assessment, testing, and research to coordinate and harmonize
20   activities, and improve data collection, analyses, curation, sharing, and warehousing.

# 7. References

Abdo N, Xia M, Kosyk O, Huang R, Sakamuru S, Austin CP, et al. (2012). The 1000 genomes toxicity screening project: Utilizing the power of human genome variation for population-scale in vitro testing. *51st Annual Meeting of the Society of Toxicology* Poster Abstract #765. http://www.toxicology.org/AI/Pub/Tox/2012Tox.pdf.

Abecasis GR, Auton A, Brooks LD, DePristo MA, Durbin RM, Handsaker RE, et al. (2012). An integrated map of genetic variation from 1,092 human genomes. *Nature* 491: 56-65. http://www.ncbi.nlm.nih.gov/pubmed/23128226.

Afridi HI, Kazi TG, Kazi N, Jamali MK, Arain MB, Jalbani N, et al. (2008). Evaluation of status of toxic metals in biological samples of diabetes mellitus patients. *Diabetes Research & Clinical Practice* 80: 280-288. http://www.ncbi.nlm.nih.gov/pubmed/18276029.

Andersen S, Pedersen KM, Bruun NH, Laurberg P. (2002). Narrow individual variations in serum T(4) and T(3) in normal subjects: A clue to the understanding of subclinical thyroid disease. *Journal of Clinical Endocrinology & Metabolism* 87: 1068-1072. http://www.ncbi.nlm.nih.gov/pubmed/11889165.

Angerer J, Bird MG, Burke TA, Doerrer NG, Needham L, Robison SH, et al. (2006). Strategic biomonitoring initiatives: moving the science forward. *Toxicological Sciences* 93: 3-10. http://www.ncbi.nlm.nih.gov/pubmed/16785253.

Ankley GT, Bennett RS, Erickson RJ, Hoff DJ, Hornung MW, Johnson RD, et al. (2010). Adverse outcome pathways: A conceptual framework to support ecotoxicology research and risk assessment. *Environmental Toxicology and Chemistry* 29: 730-741. http://service004.hpc.ncsu.edu/toxicology/websites/journalclub/linked_files/Fall10/Environ%20Toxicol%20Chem%202010%20Ankley.pdf.

Ankley GT, Gray LE. (2013). Cross-species conservation of endocrine pathways: A critical analysis of tier 1 fish and rat screening assays with 12 model chemicals. *Environmental Toxicology and Chemistry*. http://www.ncbi.nlm.nih.gov/pubmed/23401061.

Arnot JA, Mackay D TU Canadian Environmental Modelling Centre. (2007). *Risk Prioritization for a Subset of Domestic Substance List Chemicals Using the RAIDAR Model.*

Arnot JA, Mackay D, Parkerton TF, Zaleski RT, Warren CS. (2010a). Multimedia modeling of human exposure to chemical substances: The roles of food web biomagnification and biotransformation. *Environmental Toxicology and Chemistry* 29: 45-55.

Arnot JA, Mackay D, Sutcliffe R, Lo B. (2010b). Estimating farfield organic chemical exposures, intake rates and intake fractions to human age classes. *Environmental Modelling & Software* 25: 1166-1175.

ATSDR (Agency for Toxic Substances and Disease Registry). (2007). *Toxicological profile for benzene.* Retrieved from http://www.atsdr.cdc.gov/toxprofiles/tp3.pdf (accessed March 25, 2013).

Auerbach SS, Shah RR, Mav D, Smith CS, Walker NJ, Vallant MK, et al. (2010). Predicting the hepatocarcinogenic potential of alkenylbenzene flavoring agents using toxicogenomics and machine learning. *Toxicology and Applied Pharmacology* 243: 300-314. http://dx.doi.org/10.1016/j.taap.2009.11.021.

Baker M. (2010). Epigenome: Mapping in motion. *Nature Methods* 7: 181-186.

Basketter DA, Clewell H, Kimber I, Rossi A, Blaauboer B, Burrier R, et al. (2012). A roadmap for the development of alternative (non-animal) methods for systemic toxicity testing - t4 report*. *Alternatives to Animal Experiments* 29: 3-91. http://www.ncbi.nlm.nih.gov/pubmed/22307314.

Behl M, Rao D, Aagaard K, Davidson TL, Levin ED, Slotkin TA, et al. (2013). Evaluation of the association between maternal smoking, childhood obesity, and metabolic disorders: a national toxicology program workshop review. *Environmental Health Perspectives* 121: 170-180. http://www.ncbi.nlm.nih.gov/pubmed/23232494.

Bell RR, Early JL, Nonavinakere VK, Mallory Z. (1990). Effect of cadmium on blood glucose level in the rat. *Toxicology Letters* 54: 199-205. http://www.ncbi.nlm.nih.gov/pubmed/2260118.

Belson M, Kingsley B, Holmes A. (2007). Risk factors for acute leukemia in children: a review. *Environmental Health Perspectives* 115: 138-145. http://www.ncbi.nlm.nih.gov/pubmed/17366834.

Berg EL, Yang J, Melrose J, Nguyen D, Privat S, Rosler E, et al. (2010). Chemical target and pathway toxicity mechanisms defined in primary human cell systems. *Journal of Pharmacological and Toxicological Methods* 61: 3-15. http://www.ncbi.nlm.nih.gov/pubmed/19879948.

---

Bhattacharya S, Zhang Q, Carmichael PL, Boekelheide K, Andersen ME. (2011). Toxicity testing in the 21 century: defining new risk assessment approaches based on perturbation of intracellular toxicity pathways. *Public Library of Science One* 6: e20887. http://www.ncbi.nlm.nih.gov/pubmed/21701582.

Birnbaum LS. (2012). NIEHS and the future of toxicology. Presentation. *FutureTox Meeting of the Society of Toxicology* October 18-19. Arlington, VA.

Birney E. (2012). The making of ENCODE: Lessons for big-data projects. *Nature* 489: 49-51. http://www.ncbi.nlm.nih.gov/pubmed/22955613.

Blakesley V, Awni W, Locke C, Ludden T, Granneman GR, Braverman LE. (2004). Are bioequivalence studies of levothyroxine sodium formulations in euthyroid volunteers reliable? *Thyroid* 14: 191-200. http://www.ncbi.nlm.nih.gov/pubmed/15072701.

Bleicher KH, Bohm HJ, Muller K, Alanine AI. (2003). Hit and lead generation: beyond high-throughput screening. *Nature Reviews Drug Discovery* 2: 369-378. http://www.ncbi.nlm.nih.gov/entrez/query.fcgi?cmd=Retrieve&db=PubMed&dopt=Citation&list_uids=12750740.

Blount BC, Valentin-Blasini L, Osterloh JD, Mauldin JP, Pirkle JL. (2007). Perchlorate exposure of the US Population, 2001-2002. *Journal of Exposure Science and Environmental Epidemiology* 17: 400-407. http://www.ncbi.nlm.nih.gov/pubmed/17051137.

Boekelheide K, Blumberg B, Chapin RE, Cote I, Graziano JH, Janesick A, et al. (2012). Predicting later-life outcomes of early-life exposures. *Environmental Health Perspectives* 120: 1353-1361. http://dx.doi.org/10.1289/ehp.1204934.

Bollati V, Baccarelli A. (2010). Environmental epigenetics. *Heredity (Edinb)* 105: 105-112. http://www.ncbi.nlm.nih.gov/pubmed/20179736.

Borgelt C. (2013). Software for frequent pattern mining. Retrieved from http://www.borgelt.net/fpm.html (accessed March 20, 2013).

BOSC (Board of Scientific Counselors). (2010). *Review of the EPA Office of Research and Development's (ORD) Computational Toxicology Research Program (CTRP)*. Washington, DC. Retrieved from http://www.epa.gov/osp/bosc/pdf/ctox1004rpt.pdf (accessed February 26, 2013).

Bosson J, Stenfors N, Bucht A, Helleday R, Pourazar J, Holgate ST, et al. (2003). Ozone-induced bronchial epithelial cytokine expression differs between healthy and asthmatic subjects. *Clinical and Experimental Allergy* 33: 777-782. http://www.ncbi.nlm.nih.gov/pubmed/12801312.

Brown AR, Bickley LK, Le Page G, Hosken DJ, Paull GC, Hamilton PB, et al. (2011). Are toxicological responses in laboratory (inbred) zebrafish representative of those in outbred (wild) populations? - A case study with an endocrine disrupting chemical. *Environmental Science and Technology* 45: 4166-4172. http://www.ncbi.nlm.nih.gov/pubmed/21469706.

Bucher JR, Thayer K, Birnbaum LS. (2011). The Office of Health assessment and translation: A problem-solving resource for the National Toxicology Program. *Environmental Health Perspectives* 119: A196-197. http://www.ncbi.nlm.nih.gov/pubmed/21531652.

Burgoon LD. (2011). Potential of Genomic Data on PAHs to Inform Cumulative Assessment. *National Academy of Sciences Meeting: Mixtures and Cumulative Risk Assessment.* http://nas-sites.org/emergingscience/meetings/mixtures/workshop-presentations-mixtures/.

Burgoon LD, Zacharewski TR. (2008). Automated quantitative dose-response modeling and point of departure determination for large toxicogenomic and high-throughput screening data sets. *Toxicological Sciences* 104: 412-418. http://www.ncbi.nlm.nih.gov/pubmed/18441342.

Bush WS, Moore JH. (2012). Chapter 11: Genome-wide association studies. *Public Library of Science Computational Biology* 8.

Chadwick LH. (2012). The NIH roadmap epigenomics program data resource. *Epigenomics* 4: 317-324. http://www.ncbi.nlm.nih.gov/pubmed/22690667.

Chen J, Blackwell TW, Fermin D, Menon R, Chen Y, Gao J, et al. (2007). Evolutionary-conserved gene expression response profiles across mammalian tissues. *OMICS* 11: 96-115. http://www.ncbi.nlm.nih.gov/pubmed/17411398.

Chen Y, Jin JY, Mukadam S, Malhi V, Kenny JR. (2012). Application of IVIVE and PBPK modeling in prospective prediction of clinical pharmacokinetics: Strategy and approach during the drug discovery phase with four case studies. *Biopharmaceutics & Drug Disposition* 33: 85-98. http://www.ncbi.nlm.nih.gov/pubmed/22228214.

---

*This document is a draft for review purposes only and does not constitute Agency policy. Do not cite or quote.*

September 2013                                                   96

Chiu WA, Euling SY, Scott CS, Subramaniam RP. (2010). Approaches to advancing quantitative human health risk assessment of environmental chemicals in the post-genomic era. *Toxicology and Applied Pharmacology.* http://www.ncbi.nlm.nih.gov/pubmed/20353796.

Chuang HY, Lee E, Liu YT, Lee D, Ideker T. (2007). Network-based classification of breast cancer metastasis. *Molecular Systems Biology* 3: 140. http://www.ncbi.nlm.nih.gov/pubmed/17940530.

Cohen Hubal EA, Richard A, Aylward L, Edwards S, Gallagher J, Goldsmith MR, et al. (2010). Advancing exposure characterization for chemical evaluation and risk assessment. *Journal of Toxicology and Environmental Health Part B: Critical Reviews* 13: 299-313. http://www.ncbi.nlm.nih.gov/pubmed/20574904.

Collins FS. (2010). Transcript from Newshour PBS interview with Frances Collins (June 24, 2010). http://www.pbs.org/newshour/bb/science/jan-june10/genome_06-24.html.

Collins MA. (2009). Generating 'omic knowledge': The role of informatics in high content screening. *Combinatorial Chemistry & High Throughput Screening* 12: 917-925. http://www.ncbi.nlm.nih.gov/pubmed/19531005.

Cordell HJ. (2009). Detecting gene-gene interactions that underlie human diseases. *Nature Reviews Genetics* 10: 392-404. http://www.ncbi.nlm.nih.gov/pubmed/19434077.

Cortessis VK, Thomas DC, Levine AJ, Breton CV, Mack TM, Siegmund KD, et al. (2012). Environmental epigenetics: Prospects for studying epigenetic mediation of exposure-response relationships. *Human genetics* 131: 1565-1589. http://www.ncbi.nlm.nih.gov/pubmed/22740325.

Cote I, Anastas PT, Birnbaum LS, Clark RM, Dix DJ, Edwards SW, et al. (2012). Advancing the next generation of health risk assessment. *Environmental Health Perspectives* 120: 1499-1502. http://www.ncbi.nlm.nih.gov/pubmed/22875311.

Crawford-Brown D. (2013). The role of advanced biological data in the rationality of risk-based regulatory decisions. *Journal of Environmental Protection* 4: 238-249. http://www.scirp.org/journal/PaperInformation.aspx?PaperID=28638.

Crofton KM, Zoeller RT. (2005). Mode of action: neurotoxicity induced by thyroid hormone disruption during development--hearing loss resulting from exposure to PHAHs. *Critical Reviews in Toxicology* 35: 757-769. http://www.ncbi.nlm.nih.gov/pubmed/16417043.

Cronican AA, Fitz NF, Carter A, Saleem M, Shiva S, Barchowsky A, et al. (2013). Genome-wide alteration of histone H3K9 acetylation pattern in mouse offspring prenatally exposed to arsenic. *Public Library of Science One* 8: e53478. http://www.ncbi.nlm.nih.gov/pubmed/23405071.

Crump KS, Chen C, Louis TA. (2010). The future use of in vitro data in risk assessment to set human exposure standards: challenging problems and familiar solutions. *Environmental Health Perspectives* 118: 1350-1354. http://www.ncbi.nlm.nih.gov/pubmed/20562051.

De Coster S, van Larebeke N. (2012). Endocrine-disrupting chemicals: Associated disorders and mechanisms of action. *Journal of Environmental Public Health* 2012: 713696. http://www.ncbi.nlm.nih.gov/pubmed/22991565.

Dearry A. (2013). Integrating environmental health data to advance discovery. Presentation. *National Academy of Sciences Meeting: Integrating Environmental Health Data to Advance Discovery* January 10-11. http://www.youtube.com/watch?v=5TIVa2x4x7c&list=PLzsdEyVNFvgyizsegxlcIbLz1glyOIHxJ&index=2.

Degitz SJ, Holcombe GW, Flynn KM, Kosian PA, Korte JJ, Tietge JE. (2005). Progress towards development of an amphibian-based thyroid screening assay using Xenopus laevis. Organismal and thyroidal responses to the model compounds 6-propylthiouracil, methimazole, and thyroxine. *Toxicological Sciences* 87: 353-364. http://www.ncbi.nlm.nih.gov/pubmed/16002479.

Derry JM, Mangravite LM, Suver C, Furia MD, Henderson D, Schildwachter X, et al. (2012). Developing predictive molecular maps of human disease through community-based modeling. *Nature Genetics* 44: 127-130. http://www.ncbi.nlm.nih.gov/pubmed/22281773.

Devlin RB. (2012). Using Ozone to Validate a Systems Biology Approach to Toxicity Testing. *National Academy of Sciences Meeting: Systems Biology-Informed Risk Assessment* http://www.youtube.com/watch?v=5TIVa2x4x7c&list=PLzsdEyVNFvgyizsegxlcIbLz1glyOIHxJ&index=2.

Devlin RB, Duncan KE, Jardim M, Schmitt MT, Rappold AG, Diaz-Sanchez D. (2012). Controlled exposure of healthy young volunteers to ozone causes cardiovascular effects. *Circulation* 126: 104-111. http://www.ncbi.nlm.nih.gov/pubmed/22732313.

---

Diabetes Genetics Initiative of Broad Institute of Harvard, MIT LU, Novartis Institutes of BioMedical Research, Saxena R, Voight BF, Lyssenko V, et al. (2007). Genome-wide association analysis identifies loci for type 2 diabetes and triglyceride levels. *Science* 316: 1331-1336. http://www.sciencemag.org/content/316/5829/1331.abstract.

Dick E, Rajamohan D, Ronksley J, Denning C. (2010). Evaluating the utility of cardiomyocytes from human pluripotent stem cells for drug screening. *Biochemical Society Transactions* 38: 1037-1045. http://www.ncbi.nlm.nih.gov/pubmed/20659000.

Duncan KE, Crooks J, Miller DJ, Burgoon L, Schmitt MT, Edwards S, et al. (2013). Temporal profile Of gene expression alterations in primary human bronchial epithelial cells following in vivo exposure to ozone *American Thoracic Society International Conference: D105 Genetics and Epigenetics of Lung Disease* Meeting Abstract A5866. http://www.atsjournals.org/doi/abs/10.1164/ajrccm-conference.2013.187.1_MeetingAbstracts.A5866.

Duncan KE, Dailey LA, Carson JL, Hernandez ML, Peden DB, Devlin RB. (2012). Cultured basal airway epithelial cells from asthmatics display baseline gene expression profiles that differ from normal healthy cells and exhibit differential responses to ambient air pollution particles *American Journal of Respiratory and Critical Care Medicine* 185: A4291.

EC (European Commission). (2010). *Main Findings of the Report: Alternative (Non-animal) Methods for Cosmetics Testing: Current Status and Future Prospects.* Institute for Health and Consumer Protection. Retrieved from http://www.iivs.org/workspace/assets/news-assets/findings_cosmetics_2011.pdf (accessed February 22, 2013).

EC (European Commission). (2011). *4th Report on the Implementation of the "Community Strategy for Endocrine Disruptors" a Range of Substances Suspected of Interfering with the Hormone Systems of Humans and Wildlife.* (COM (1999) 706). Brussels. Retrieved from http://ec.europa.eu/environment/endocrine/documents/sec_2011_1001_en.pdf (accessed March 4, 2013).

ECHA (European Chemicals Agency). (2013a). *Evaluation Under REACH: Progress Report 2012.* Helsinki. Retrieved from http://echa.europa.eu/documents/10162/13628/evaluation_report_2012_en.pdf (accessed March 29, 2013).

ECHA (European Chemicals Agency). (2013b). *Proposal for Identification of a Substance as a CMR 1A or 1B, Pbt, vPvB or a Substance of an Equivalent Level of Concern.* Retrieved from http://echa.europa.eu/documents/10162/13638/annex_xv_svhc_4nonylphenol_en.pdf (accessed March 28, 2013).

Ein-Dor L, Kela I, Getz G, Givol D, Domany E. (2005). Outcome signature genes in breast cancer: Is there a unique set? *Bioinformatics* 21: 171-178. http://www.ncbi.nlm.nih.gov/pubmed/15308542.

Eisenberg M, Samuels M, DiStefano JJ, 3rd. (2008). Extensions, validation, and clinical applications of a feedback control system simulator of the hypothalamo-pituitary-thyroid axis. *Thyroid* 18: 1071-1085. http://www.ncbi.nlm.nih.gov/pubmed/18844475.

El Muayed M, Raja MR, Zhang X, MacRenaris KW, Bhatt S, Chen X, et al. (2012). Accumulation of cadmium in insulin-producing β cells. *Islets* 4: 405-416.

Ellinger-Ziegelbauer H, Gmuender H, Bandenburg A, Ahr HJ. (2008). Prediction of a carcinogenic potential of rat hepatocarcinogens using toxicogenomics analysis of short-term in vivo studies. *Mutation Research* 637: 23-39. http://dx.doi.org/10.1016/j.mrfmmm.2007.06.010.

Emdin SO, Dodson GG, Cutfield JM, Cutfield SM. (1980). Role of zinc in insulin biosynthesis. Some possible zinc-insulin interactions in the pancreatic B-cell. *Diabetologia* 19: 174-182. http://www.ncbi.nlm.nih.gov/pubmed/6997118.

ENCODE Project Consortium. (2012). An integrated encyclopedia of DNA elements in the human genome. *Nature* 489: 57-74. http://www.ncbi.nlm.nih.gov/pubmed/22955616.

EPA (U.S. Environmental Protection Agency). (1995). *The Use of the Benchmark Dose Approach in Health Risk Assessment.* (EPA/630/R-94/007). Washington, DC. Retrieved from http://www.epa.gov/raf/publications/pdfs/BENCHMARK.PDF (accessed March 8, 2013).

EPA (U.S. Environmental Protection Agency). (2000). Benzene (CASRN 71-43-2) IRIS Summary. II. Carcinogenicity Assessment for Lifetime Exposure. January 19. Retrieved from http://www.epa.gov/iris/subst/0276.htm (accessed March 25, 2013).

EPA (U.S. Environmental Protection Agency). (2008). *Uncertainty and Variability in Physiologically Based Pharmacokinetic Models: Key Issues and Case Studies.* Washington, DC. Retrieved from http://ofmpub.epa.gov/eims/eimscomm.getfile?p_download_id=477286.

*This document is a draft for review purposes only and does not constitute Agency policy. Do not cite or quote.*

September 2013                                                                 98

EPA (U.S. Environmental Protection Agency). (2009a). *Strategic Plan for the Future of Toxicity Testing and Risk Assessment at the U.S. EPA*. Washington, DC: Office of the Science Advisor, Science Policy Council. Retrieved from http://www.epa.gov/spc/toxicitytesting/docs/toxtest_strategy_032309.pdf (accessed February 22, 2013).

EPA (U.S. Environmental Protection Agency). (2009b). Virtual Tissues Project. April 21-22. Retrieved from http://www.epa.gov/ncct/virtual_tissues/ (accessed April 2, 2013).

EPA (U.S. Environmental Protection Agency). (2010). *Advancing the Next Generation (NexGen) of Risk Assessment: The Prototypes Workshop*. Research Triangle Park, NC. Retrieved from http://www.epa.gov/risk/nexgen/docs/NexGen-Prototypes-Workshop-Summary.pdf (accessed February 22, 2013).

EPA (U.S. Environmental Protection Agency). (2011a). *Advancing the Next Generation (NexGen) of Risk Assessment: Public Dialogue Conference*. Washington, DC. Retrieved from http://www.epa.gov/risk/nexgen/docs/NexGen-Public-Conf-Summary.pdf (accessed February 22, 2013).

EPA (U.S. Environmental Protection Agency). (2011b). *Framework for an EPA Chemical Safety for Sustainability Research Program*. Washington, DC: Office of Research and Development. Retrieved from http://www.epa.gov/ord/priorities/docs/CSSFramework.pdf (accessed February 22, 2013).

EPA (U.S. Environmental Protection Agency). (2011c). *The Incorporation of In Silico Models and In Vitro High Throughput Assays in the Endocrine Disruptor Screening Program (EDSP) for Prioritization and Screening*. Washington, DC. Retrieved from http://www.epa.gov/endo/pubs/edsp21_work_plan_summary%20_overview_final.pdf (accessed February 22, 2013).

EPA (U.S. Environmental Protection Agency). (2011d). *Integrated Science Assessment of Ozone and Related Photochemical Oxidants (Final Report)*. Washington, DC. Retrieved from http://ctpub.cpa.gov/ncca/isa/recordisplay.cfm?deid=247492 (accessed March 28, 2013).

EPA (U.S. Environmental Protection Agency). (2011e). *Use of "Omic" Technology to Inform the Risk Assessment, Support Document for Case Study: Propiconazole. Appendices A and B*. Washington, DC: Federal Insecticide, Fungicide, and Rodenticide Act Scientific Advisory Panel (FIFRA SAP). Retrieved from http://www.regulations.gov/#!documentDetail;D=EPA-HQ-OPP-2011-0284-0003 (accessed February 22, 2013).

EPA (U.S. Environmental Protection Agency). (2012a). Advancing the Next Generation (NexGen) of Risk Assessment. July 31. Retrieved from http://www.epa.gov/risk/nexgen/ (accessed March 21, 2013).

EPA (U.S. Environmental Protection Agency). (2012b). Basic Information About Benzene in Drinking Water. Retrieved from http://water.epa.gov/drink/contaminants/basicinformation/benzene.cfm (accessed March 25, 2013).

EPA (U.S. Environmental Protection Agency). (2012c). Computational Toxicology Research. Retrieved from http://www.epa.gov/ncct/ (accessed February 22, 2013).

EPA (U.S. Environmental Protection Agency). (2012d). *Computational Toxicology Research: Overview Materials for EPA Science Advisory Board Exposure & Human Health Committee*. Washington, DC. Retrieved from http://yosemite.epa.gov/sab/sabproduct.nsf/AEACDCDADEDFAAF3852579F2005A4C5A/$File/EPA+CompTox+SAB+Materials+4-26-2012+v4.pdf (accessed February 22, 2013).

EPA (U.S. Environmental Protection Agency). (2012e). Endocrine Disruptor Screening Program (EDSP). November 27. Retrieved from http://www.epa.gov/endo/ (accessed March 4, 2013).

EPA (U.S. Environmental Protection Agency). (2013). Sustainable futures: Models & methods. February 6. Retrieved from http://www.epa.gov/oppt/sf/tools/methods.htm (accessed March 22, 2013).

Eskenazi D, Bradman A, Gladstone EA, Jaramillo S, Birch K, Holland N. (2003). CHAMACOS, A longitudinal birth cohort study. Lessons from the fields. *Journal of Children's Health* 1: 3-27. http://ehs.sph.berkeley.edu/holland/publications/files/Eskenazi2003.pdf

Fang X, Bai C, Wang X. (2012). Bioinformatics insights into acute lung injury/acute respiratory distress syndrome. *Clinical and Translational Medicine* 1: 9. http://www.ncbi.nlm.nih.gov/pubmed/23369517.

Fielden MR, Adai A, Dunn RT, Olaharski A, Searfoss G, Sina J, et al. (2011). Development and evaluation of a genomic signature for the prediction and mechanistic assessment of nongenotoxic hepatocarcinogens in the rat. *Toxicological Sciences* 124: 54-74. http://dx.doi.org/10.1093/toxsci/kfr202.

---

Fielden MR, Brennan R, Gollub J. (2007). A gene expression biomarker provides early prediction and mechanistic assessment of hepatic tumor induction by nongenotoxic chemicals. *Toxicological Sciences* 99: 90-100. http://dx.doi.org/10.1093/toxsci/kfm156.

Fielden MR, Eynon BP, Natsoulis G, Jarnagin K, Banas D, Kolaja KL. (2005). A gene expression signature that predicts the future onset of drug-induced renal tubular toxicity. *Toxicologic Pathology* 33: 675-683. http://dx.doi.org/10.1080/01926230500321213.

Fielden MR, Nie A, McMillian M, Elangbam CS, Trela BA, Yang Y, et al. (2008). Interlaboratory evaluation of genomic signatures for predicting carcinogenicity in the rat. *Toxicological Sciences* 103: 28-34. http://dx.doi.org/10.1093/toxsci/kfn022.

Fortunel NO, Otu HH, Ng HH, Chen J, Mu X, Chevassut T, et al. (2003). Comment on "'Stemness': Transcriptional profiling of embryonic and adult stem cells" and "a stem cell molecular signature". *Science* 302: 393; author reply 393. http://www.ncbi.nlm.nih.gov/pubmed/14563990.

Fowler BA, Whittaker MH, Lipsky M, Wang G, Chen XQ. (2004). Oxidative stress induced by lead, cadmium and arsenic mixtures: 30-day, 90-day, and 180-day drinking water studies in rats: an overview. *Biometals* 17: 567-568. http://www.ncbi.nlm.nih.gov/pubmed/15688865.

Friend S. (2013). Scientific opportunities from heterogeneous biological data analysis: Overcoming complexity. Presentation. *National Academy of Sciences Meeting: Integrating Environmental Health Data to Advance Discovery* January 10-11. http://nas-sites.org/emergingscience/files/2013/01/FRIEND-Jan10-final.pdf.

Froehlicher M, Liedtke A, Groh KJ, Neuhauss SC, Segner H, Eggen RI. (2009). Zebrafish (Danio rerio) neuromast: Promising biological endpoint linking developmental and toxicological studies. *Aquatic Toxicology* 95: 307-319. http://www.ncbi.nlm.nih.gov/pubmed/19467721.

Gangwal S, Reif DM, Mosher S, Egeghy PP, Wambaugh JF, Judson RS, et al. (2012). Incorporating exposure information into the toxicological prioritization index decision support framework. *Science of the Total Environment* 435-436: 316-325. http://www.ncbi.nlm.nih.gov/pubmed/22863807.

Garte S, Taioli E, Popov T, Bolognesi C, Farmer P, Merlo F. (2008). Genetic susceptibility to benzene toxicity in humans. *Journal of Toxicology and Environmental Health, Part A* 71: 1482-1489. http://www.ncbi.nlm.nih.gov/pubmed/18836923.

George BJ, Schultz BD, Palma T, Vette AF, Whitaker DA, Williams RW. (2011). *An Evaluation of EPA's National-Scale Air Toxics Assessment (NATA): Comparison with Benzene Measurements in Detroit, Michigan.* U.S. Environmental Protection Agency. Retrieved from http://www.google.com/url?sa=t&rct=j&q=an%20evaluation%20of%20epa's%20national-scale%20air%20toxics%20assessment&source=web&cd=2&ved=0CEEQFjAB&url=http%3A%2F%2Fcfpub.epa.gov%2Fsi%2Fsi_public_file_download.cfm%3Fp_download_id%3D500973&ei=zYEmUfj7L62r2AWMuoC4BQ&usg=AFQjCNGLT3m4187J8hvEnaY_fF6c24-uMA&bvm=bv.42661473,d.b2I (accessed February 22, 2013).

German Federal Ministry for Education and Research. (2013). Virtual Liver Network. Retrieved from http://www.virtual-liver.de/wordpress/en (accessed March 28, 2013).

Gibbs-Flournoy EA, Simmons SO, Bromberg PA, Dick TP, Samet JM. (2013). Monitoring intracellular redox changes in ozone-exposed airway epithelial cells. *Environmental Health Perspectives* 121: 312-317. http://www.ncbi.nlm.nih.gov/pubmed/23249900.

Gibson GG, Rostami-Hodjegan A. (2007). Modelling and simulation in prediction of human xenobiotic absorption, distribution, metabolism and excretion (ADME): In vitro-in vivo extrapolations (IVIVE). *Xenobiotica* 37: 1013-1014. http://www.ncbi.nlm.nih.gov/pubmed/17968734.

Godderis L, Thomas R, Hubbard AE, Tabish AM, Hoet P, Zhang L, et al. (2012). Effect of chemical mutagens and carcinogens on gene expression profiles in human TK6 cells. *Public Library of Science One* 7: e39205. http://europepmc.org/abstract/MED/22723965.

Golbraikh A, Wang XS, Zhu H, Tropsha A. (2012). Predictive QSAR modeling: Methods and applications in drug discovery and chemical risk assessment. *Handbook of Computational Chemistry* 1309-1342. http://link.springer.com/referenceworkentry/10.1007%2F978-94-007-0711-5_37#.

Gold LS, Manley NB, Slone TH, Ward JM. (2001). Compendium of chemical carcinogens by target organ: Results of chronic bioassays in rats, mice, hamsters, dogs, and monkeys. *Toxicologic Pathology* 29: 639-652.

---

Goldsmith MR, Peterson SD, Chang DT, Transue TR, Tornero-Velez R, Tan YM, et al. (2012). Informing mechanistic toxicology with computational molecular models. *Methods in Molecular Biology* 929: 139-165. http://www.ncbi.nlm.nih.gov/pubmed/23007429.

Goldstein BD. (1988). Benzene toxicity. *Occupational Medicine* 3: 541-554. http://www.ncbi.nlm.nih.gov/pubmed/3043738.

Greenawalt DM, Sieberts SK, Cornelis MC, Girman CJ, Zhong H, Yang X, et al. (2012). Integrating genetic association, genetics of gene expression, and single nucleotide polymorphism set analysis to identify susceptibility Loci for type 2 diabetes mellitus. *American Journal of Epidemiology* 176: 423-430. http://www.ncbi.nlm.nih.gov/pubmed/22865700.

Guryev V, Koudijs MJ, Berezikov E, Johnson SL, Plasterk RH, van Eeden FJ, et al. (2006). Genetic variation in the zebrafish. *Genome Research* 16: 491-497. http://www.ncbi.nlm.nih.gov/pubmed/16533913.

Hansen KD, Timp W, Bravo HC, Sabunciyan S, Langmead B, McDonald OG, et al. (2011). Increased methylation variation in epigenetic domains across cancer types. *Nature Genetics* 43: 768-775. http://www.ncbi.nlm.nih.gov/pubmed/21706001.

Harrill AH, Desmet KD, Wolf KK, Bridges AS, Eaddy JS, Kurtz CL, et al. (2012). A mouse diversity panel approach reveals the potential for clinical kidney injury due to DB289 not predicted by classical rodent models. *Toxicological Sciences* 130: 416-426. http://www.ncbi.nlm.nih.gov/pubmed/22940726.

Harrill AH, Ross PK, Gatti DM, Threadgill DW, Rusyn I. (2009). Population-based discovery of toxicogenomics biomarkers for hepatotoxicity using a laboratory strain diversity panel. *Toxicological Sciences* 110. 235-243. http://www.ncbi.nlm.nih.gov/pubmed/19420014.

Hatch GE, Slade R, Harris LP, McDonnell WF, Devlin RB, Koren HS, et al. (1994). Ozone dose and effect in humans and rats. A comparison using oxygen-18 labeling and bronchoalveolar lavage. *American Journal of Respiratory and Critical Care Medicine* 150: 676-683. http://www.ncbi.nlm.nih.gov/pubmed/8087337.

Hatzimichael E, Crook T. (2013). Cancer epigenetics: new therapies and new challenges. *Journal of Drug Delivery* 2013: 529312. http://www.ncbi.nlm.nih.gov/pubmed/23533770.

Hester RL, Brown AJ, Husband L, Illiescu R, Pruett D, Summers R, et al. (2011). HumMod: A modeling environment for the simulation of integrative human physiology. *Frontiers in Physiology* 2.

Holzhutter HG, Drasdo D, Preusser T, Lippert J, Henney AM. (2012). The virtual liver: A multidisciplinary, multilevel challenge for systems biology. *Wiley Interdisciplinary Reviews: Systems Biology and Medicine* 4: 221-235. http://www.ncbi.nlm.nih.gov/pubmed/22246674.

Houck KA, Kavlock RJ. (2008). Understanding mechanisms of toxicity: Insights from drug discovery research. *Toxicology and Applied Pharmacology* 227: 163-178. http://www.ncbi.nlm.nih.gov/pubmed/18063003.

Howlader N, Noone AM, Krapcho M, Garshell J, Neyman N, Altekruse SF, et al. (2013). *SEER Cancer Statistics Review, 1975-2010*. Bethesda, MD: U.S. National Institutes of Health, National Cancer Institute. Retrieved from http://seer.cancer.gov/csr/1975_2010/, based on November 2012 SEER data submission, posted to the SEER web site, April 2013.

Hubal EA. (2009). Biologically relevant exposure science for 21st century toxicity testing. *Toxicological Sciences* 111: 226-232. http://www.ncbi.nlm.nih.gov/pubmed/19602574.

Hunter P, Coveney PV, de Bono B, Diaz V, Fenner J, Frangi AF, et al. (2010). A vision and strategy for the virtual physiological human in 2010 and beyond. *Philosophical Transactions Series A: Mathematical, Physical, and Engineering Sciences* 368: 2595-2614. http://www.ncbi.nlm.nih.gov/pubmed/20439264.

Hunter P, Robbins P, Noble D. (2002). The IUPS human Physiome Project. *Pflugers Archiv* 445: 1-9. http://www.ncbi.nlm.nih.gov/pubmed/12397380.

Hurley PM. (1998). Mode of carcinogenic action of pesticides inducing thyroid follicular cell tumors in rodents. *Environmental Health Perspectives* 106: 437-445. http://www.ncbi.nlm.nih.gov/pubmed/9681970.

IARC (IARc Cancer). (2012). *Monograph 100F: Benzene*. Retrieved from http://monographs.iarc.fr/ENG/Monographs/vol100F/mono100F-24.pdf (accessed March 25, 2013).

Ilhan G, Karakus S, Andic N. (2006). Risk factors and primary prevention of acute leukemia. *Asian Pacific Journal of Cancer Prevention* 7: 515-517. http://www.ncbi.nlm.nih.gov/pubmed/17250419.

Ingelman-Sundberg M. (2005). Genetic polymorphisms of cytochrome P450 2D6 (CYP2D6): clinical consequences, evolutionary aspects and functional diversity. *Pharmacogenomics* 5: 6-13. http://www.ncbi.nlm.nih.gov/pubmed/15492763.

Irons RD, Chen Y, Wang X, Ryder J, Kerzic PJ. (2013). Acute myeloid leukemia following exposure to benzene more closely resembles de novo than therapy related-disease. *Genes Chromosomes Cancer* 52: 887-894. http://www.ncbi.nlm.nih.gov/pubmed/23840003.

Jack J, Wambaugh JF, Shah I. (2011). Simulating quantitative cellular responses using asynchronous threshold Boolean network ensembles. *BMC Systems Biology* 5: 109. http://www.ncbi.nlm.nih.gov/pubmed/21745399.

Jiang X, Kumar K, Hu X, Wallqvist A, Reifman J. (2008). DOVIS 2.0: An efficient and easy to use parallel virtual screening tool based on AutoDock 4.0. *Chemical Central Journal* 2: 18. http://www.ncbi.nlm.nih.gov/pubmed/18778471.

Jubeaux G, Audouard-Combe F, Simon R, Tutundjian R, Salvador A, Geffard O, et al. (2012). Vitellogenin-like proteins among invertebrate species diversity: potential of proteomic mass spectrometry for biomarker development. *Environmental Science and Technology* 46: 6315-6323. http://www.ncbi.nlm.nih.gov/pubmed/22578134.

Judson RS, Kavlock R, Martin M, Reif D, Houck K, Knudsen T, et al. (2013). Perspectives on validation of high-throughput assays supporting 21st century toxicity testing. *Alternatives to Animal Experiments* 30: 51-56. http://www.ncbi.nlm.nih.gov/pubmed/23338806.

Judson RS, Kavlock RJ, Setzer RW, Hubal EA, Martin MT, Knudsen TB, et al. (2011). Estimating toxicity-related biological pathway altering doses for high-throughput chemical risk assessment. *Chemical Research in Toxicology* 24: 451-462. http://dx.doi.org/10.1021/tx100428e.

Judson RS, Martin MT, Reif DM, Houck KA, Knudsen TB, Rotroff DM, et al. (2010). Analysis of eight oil spill dispersants using rapid, in vitro tests for endocrine and other biological activity. *Environmental Science and Technology* 44: 5979-5985. http://dx.doi.org/10.1021/es102150z.

Kanehisa Laboratories. (2013). KEGG: Kyoto encyclopedia of genes and genomes. Retrieved from http://www.genome.jp/kegg/ (accessed February 22, 2013).

Kavlock R, Chandler K, Houck K, Hunter S, Judson R, Kleinstreuer N, et al. (2012). Update on EPA's ToxCast program: providing high throughput decision support tools for chemical risk management. *Chemical Research in Toxicology* 25: 1287-1302. http://www.ncbi.nlm.nih.gov/pubmed/22519603.

Kim CS, Alexis NE, Rappold AG, Kehrl H, Hazucha MJ, Lay JC, et al. (2011). Lung function and inflammatory responses in healthy young adults exposed to 0.06 ppm ozone for 6.6 hours. *American Journal of Respiratory and Critical Care Medicine* 183: 1215-1221. http://www.ncbi.nlm.nih.gov/pubmed/21216881.

Knudsen T, Daston GP. (2010). Virtual tissues and developmental systems biology (Chapter 23). In T Knudsen, G Daston (Eds.), *Second Edition Comprehensive Toxicology* (pp. 347-358). Oxford, UK: Elsevier Ltd.

Knudsen T, DeWoskin RS. (2011). Systems modeling in development toxicity. In *Handbook of Systems Toxicology*: John Wiley & Sons, Ltd.

Knudsen T, Kavlock RJ, Daston GP, Stedman D, Hixon M, Kim JH. (2011). Developmental toxicity testing for safety assessment: New approaches and technologies. *Birth Defects Research: Part B, Developmental and Reproductive Toxicology* 92: 413-420. http://www.ncbi.nlm.nih.gov/pubmed/21770025.

Kondraganti SR, Fernandez-Salguero P, Gonzalez FJ, Ramos KS, Jiang W, Moorthy B. (2003). Polycyclic aromatic hydrocarbon-inducible DNA adducts: Evidence by 32P-postlabeling and use of knockout mice for Ah receptor-independent mechanisms of metabolic activation in vivo. *International Journal of Cancer* 103: 5-11. http://www.ncbi.nlm.nih.gov/pubmed/12455047.

Koturbash I, Beland FA, Pogribny IP. (2011). Role of epigenetic events in chemical carcinogenesis--a justification for incorporating epigenetic evaluations in cancer risk assessment. *Toxicology Mechanisms and Methods* 21: 289-297. http://www.ncbi.nlm.nih.gov/pubmed/21495867.

Krauth D, Woodruff TJ, Bero L. (2013). Instruments for assessing risk of bias and other methodological criteria of published animal studies: a systematic review. *Environmental Health Perspectives* 121: 985-992. http://www.ncbi.nlm.nih.gov/pubmed/23771496.

Krewski D, Hogan V, Turner MC, Zeman PL, McDowell I, Edwards N, et al. (2007). An integrated framework for risk management and population health. *Human and Ecological Risk Assessment* 13: 1288-1312.

*This document is a draft for review purposes only and does not constitute Agency policy. Do not cite or quote.*

September 2013                                                                102

Krewski D, Westphal M, Paoli G, Croteau MC, Al-Zoughool M, Andersen ME, et al. (2013). A framework for the next generation of risk science.

Lan Q, Zhang L, Li G, Vermeulen R, Weinberg RS, Dosemeci M, et al. (2004). Hematotoxicity in workers exposed to low levels of benzene. *Science* 306: 1774-1776. http://www.ncbi.nlm.nih.gov/pubmed/15576619.

Lau FK, Decastro BR, Mills-Herring L, Tao L, Valentin-Blasini L, Alwis KU, et al. (2013). Urinary perchlorate as a measure of dietary and drinking water exposure in a representative sample of the United States population 2001-2008. *Journal of Exposure Science and Environmental Epidemiology* 23: 207-214. http://www.ncbi.nlm.nih.gov/pubmed/23188482.

Lock EF, Abdo N, Huang R, Xia M, Kosyk O, O'Shea SH, et al. (2012). Quantitative high-throughput screening for chemical toxicity in a population-based in vitro model. *Toxicological Sciences* 126: 578-588. http://dx.doi.org/10.1093/toxsci/kfs023.

Lossos IS, Czerwinski DK, Alizadeh AA, Wechser MA, Tibshirani R, Botstein D, et al. (2004). Prediction of survival in diffuse large-B-cell lymphoma based on the expression of six genes. *New England Journal of Medicine* 350: 1828-1837. http://www.ncbi.nlm.nih.gov/pubmed/15115829.

Lvovs D, Favorova OO, Favorov AV. (2012). A polygenic approach to the study of polygenic diseases. *Acta Naturae* 4: 59-71. http://www.ncbi.nlm.nih.gov/pubmed/23150804.

Makris SL, Kim JH, Ellis A, Faber W, Harrouk W, Lewis JM, et al. (2011). Current and future needs for developmental toxicity testing. *Birth Defects Research: Part B, Developmental and Reproductive Toxicology* 92: 384-394. http://www.ncbi.nlm.nih.gov/pubmed/21922641.

Martin MT, Knudsen TB, Reif DM, Houck KA, Judson RS, Kavlock RJ, et al. (2011). Predictive Model of Rat Reproductive Toxicity from ToxCast High Throughput Screening. *Biology of Reproduction* 85: 327-339. http://www.ncbi.nlm.nih.gov/pubmed/21565999.

Maull EA, Ahsan H, Edwards J, Longnecker MP, Navas-Acien A, Pi J, et al. (2012). Evaluation of the association between arsenic and diabetes: a National Toxicology Program workshop review. *Environmental Health Perspectives* 120: 1658-1670. http://www.ncbi.nlm.nih.gov/pubmed/22889723.

Mayr LM, Bojanic D. (2009). Novel trends in high-throughput screening. *Current Opinion in Pharmacology* 9: 580-588. http://www.ncbi.nlm.nih.gov/entrez/query.fcgi?cmd=Retrieve&db=PubMed&dopt=Citation&list_uids=19775937.

McConnell ER, Bell SM. (2013). Systematic Omics Analysis Review (SOAR) tool to support risk assessment. *Submitted*.

McDonnell WF, Stewart PW, Smith MV, Kim CS, Schelegle ES. (2012). Prediction of lung function response for populations exposed to a wide range of ozone conditions. *Inhalation Toxicology* 24: 619-633. http://www.ncbi.nlm.nih.gov/pubmed/22906168.

McDougall CM, Blaylock MG, Douglas JG, Brooker RJ, Helms PJ, Walsh GM. (2008). Nasal epithelial cells as surrogates for bronchial epithelial cells in airway inflammation studies. *American Journal of Respiratory Cell and Molecular Biology* 39: 560-568. http://www.ncbi.nlm.nih.gov/pubmed/18483420.

McHale CM, Zhang L, Lan Q, Vermeulen R, Li G, Hubbard AE, et al. (2011). Global gene expression profiling of a population exposed to a range of benzene levels. *Environmental Health Perspectives* 119: 628-634. http://dx.doi.org/10.1289/ehp.1002546.

McHale CM, Zhang L, Smith MT. (2012). Current understanding of the mechanism of benzene-induced leukemia in humans: Implications for risk assessment. *Carcinogenesis* 33: 240-252.

Mechanic LE, Chen HS, Amos CI, Chatterjee N, Cox NJ, Divi RL, et al. (2012). Next generation analytic tools for large scale genetic epidemiology studies of complex diseases. *Genetic epidemiology* 36: 22-35. http://www.ncbi.nlm.nih.gov/pubmed/22147673.

Medzhitov R. (2008). Origin and physiological roles of inflammation. *Nature* 454: 428-435. http://www.ncbi.nlm.nih.gov/pubmed/18650913.

Meissner A. (2012). What can epigenomics do for you? *Genome Biology* 13: 420. http://www.ncbi.nlm.nih.gov/pubmed/23095436.

Miao X, Sun W, Fu Y, Miao L, Cai L. (2013). Zinc homeostasis in the metabolic syndrome and diabetes. *Frontiers of Medicine* 7: 31-52. http://www.ncbi.nlm.nih.gov/pubmed/23385610.

Miller MD, Crofton KM, Rice DC, Zoeller RT. (2009). Thyroid-disrupting chemicals: Interpreting upstream biomarkers of adverse outcomes. *Environmental Health Perspectives* 117: 1033-1041. http://www.ncbi.nlm.nih.gov/pubmed/19654909.

Morales-Ruan Mdel C, Villalpando S, Garcia-Guerra A, Shamah-Levy T, Robledo-Perez R, Avila-Arcos MA, et al. (2012). Iron, zinc, copper and magnesium nutritional status in Mexican children aged 1 to 11 years. *Salud Publica de Mexico* 54: 125-134. http://www.ncbi.nlm.nih.gov/pubmed/22535171.

Moraru, II, Schaff JC, Slepchenko BM, Blinov ML, Morgan F, Lakshminarayana A, et al. (2008). Virtual cell modelling and simulation software environment. *IET Systems Biology* 2: 352-362. http://www.ncbi.nlm.nih.gov/pubmed/19045830.

Mortensen HM, Euling SY. (2013). Integrating mechanistic and polymorphism data to characterize human genetic susceptibility for environmental chemical risk assessment in the 21st century. *Toxicology and Applied Pharmacology* 271: 395-404. http://dx.doi.org/10.1016/j.taap.2011.01.015.

Murk AJ, Rijntjes E, Blaauboer BJ, Clewell R, Crofton KM, Dingemans MM, et al. (2013). Mechanism-based testing strategy using in vitro approaches for identification of thyroid hormone disrupting chemicals. *Toxicology In Vitro*. http://www.ncbi.nlm.nih.gov/pubmed/23453986.

NAS (National Academy of Sciences). (2007). *Scientific Review of the Proposed Risk Assessment Bulletin from the Office of Management and Budget.* (9780309104777). Washington, DC: The National Academies Press. Retrieved from http://www.nap.edu/openbook.php?record_id=11811.

NCBI (National Center for Biotechnology Information). (2009). Epigenomics (database). Retrieved from http://www.ncbi.nlm.nih.gov/epigenomics (accessed March 4, 2013).

NCBI (National Center for Biotechnology Information). (2012a). Gene Expression Omnibus. Retrieved from http://www.ncbi.nlm.nih.gov/geo/ (accessed February 22, 2013).

NCBI (National Center for Biotechnology Information). (2012b). *Reference SNP(refSNP) Cluster Report: rs13266634.* dbSNP: Short Genetic Variations. Retrieved from http://www.ncbi.nlm.nih.gov/projects/SNP/snp_ref.cgi?rs=13266634 (accessed March 20, 2013).

NCBI (National Center for Biotechnology Information). (2013). PubMed Website. Retrieved from http://www.ncbi.nlm.nih.gov/pubmed (accessed April 2, 2013).

NHGRI (National Human Genome Research Institute). (2012). Home Page. Retrieved from http://www.genome.gov/ (accessed February 22, 2013).

NHGRI (National Human Genome Research Institute). (2013). A Catalog of Published Genome-Wide Association Studies. Retrieved from http://www.genome.gov/gwastudies/ (accessed March 25, 2013).

Nichols JW, Breen M, Denver RJ, Distefano JJ, 3rd, Edwards JS, Hoke RA, et al. (2011). Predicting chemical impacts on vertebrate endocrine systems. *Environmental Toxicology and Chemistry* 30: 39-51. http://www.ncbi.nlm.nih.gov/pubmed/20963851.

Nie AY, Mcmillian M, Parker JB, Leone A, Bryant S, Yieh L, et al. (2006). Predictive toxicogenomics approaches reveal underlying molecular mechanisms of nongenotoxic carcinogenicity. *Molecular Carcinogenesis* 45: 914-933.

NIEHS (National Institute of Environmental Health Sciences). (2012a). Host Susceptibility Program. September 7. Retrieved from http://ntp.niehs.nih.gov/?objectid=B76D131B-F1F6-975E-71657BC3DC88C299 (accessed April 2, 2013).

NIEHS (National Institute of Environmental Health Sciences). (2012b). NIEHS SNPs Environmental Genome Project. Retrieved from http://egp.gs.washington.edu/ (accessed March 25, 2013).

NIEHS (National Institute of Environmental Health Sciences). (2012c). NIEHS Strategic Plan. Retrieved from http://niehs.nih.gov/about/strategicplan/index.cfm (accessed February 22, 2013).

NIEHS (National Institute of Environmental Health Sciences). (2013). Comparative Toxicogenomic Database (CTD)™. February 6. Retrieved from http://ctdbase.org/ (accessed March 4, 2013).

NIH (National Institutes of Health). (2012). NIH Chemical Genomics Center. Retrieved from http://www.ncats.nih.gov/research/reengineering/ncgc/ncgc.html (accessed February 22, 2013).

NIOSH (National Institute of Occupational Safety and Health). (1992). *Recommendations for Occupational Safety and Health: Compendium of Policy Documents and Statements.* Retrieved from http://www.cdc.gov/niosh/pdfs/92-100.pdf (accessed March 25, 2013).

North M, Tandon VJ, Thomas R, Loguinov A, Gerlovina I, Hubbard AE, et al. (2011). Genome-wide functional profiling reveals genes required for tolerance to benzene metabolites in yeast. *Public Library of Science One* 6: e24205. http://www.ncbi.nlm.nih.gov/pubmed/21912624.

NRC (National Research Council). (2006). *Human Biomonitoring for Environmental Chemicals.* The National Academies Press. Retrieved from http://www.nap.edu/openbook.php?record_id=11700.

NRC (National Research Council). (2007). *Toxicity Testing in the 21st Century: A Vision and a Strategy.* Washington DC. Retrieved from http://dels.nas.edu/resources/static-assets/materials-based-on-reports/reports-in-brief/Toxicity_Testing_final.pdf (accessed February 22, 2013).

NRC (National Research Council). (2009). *Science and Decisions: Advancing Risk Assessment.* Washington, DC. Retrieved from http://www.nap.edu/catalog/12209.html (accessed February 22, 2013).

NRC (National Research Council). (2011). *Predicting Later-Life Outcomes of Early-Life Exposures.* Retrieved from http://nas-sites.org/emergingscience/files/2011/05/inutero_final_April2011.pdf (accessed April 2, 2013).

O'Shea SH, Schwarz J, Kosyk O, Ross PK, Ha MJ, Wright FA, et al. (2011). In vitro screening for population variability in chemical toxicity. *Toxicological Sciences* 119: 398-407. http://www.ncbi.nlm.nih.gov/pubmed/20952501.

OECD (Organization for Economic Cooperation and Development). (2004). *OECD Principles for the Validation, for Regulatory Purposes, of Quantitative Structure Activity Relationship Models.* Paris, France. Retrieved from http://www.oecd.org/chemicalsafety/assessmentofchemicals/37849783.pdf (accessed February 22, 2013).

OECD (Organization for Economic Cooperation and Development). (2012). The OECD QSAR Toolbox. Version 3.0. Retrieved from http://www.oecd.org/env/chemicalsafetyandbiosafety/assessmentofchemicals/theoecdqsartoolbox.htm (accessed February 22, 2013).

OMIM (Online Mendelian Inheritance in Man). (2012). Solute Carrier Family 30 (Zinc Transporter), Member 8, SLC30A8 November 13. Retrieved from http://omim.org/entry/611145#0001 (accessed March 21, 2013).

Oracle. (2013a). 5 Configuring Rule Sets. Retrieved from http://docs.oracle.com/cd/B28359_01/server.111/b31222/cfrulset.htm#DVADM70150 (accessed March 20, 2013).

Oracle. (2013b). Glossary: "Lift". Retrieved from http://docs.oracle.com/cd/B28359_01/datamine.111/b28129/glossary.htm (accessed March 20, 2013).

Padilla S, Corum D, Padnos B, Hunter DL, Beam A, Houck KA, et al. (2012). Zebrafish developmental screening of the ToxCast Phase I chemical library. *Reproductive Toxicology* 33: 174-187. http://www.ncbi.nlm.nih.gov/pubmed/22182468.

Pare G, Chasman DI, Parker AN, Nathan DM, Miletich JP, Zee RY, et al. (2008). Novel association of HK1 with glycated hemoglobin in a non-diabetic population: A genome-wide evaluation of 14,618 participants in the Women's Genome Health Study. *Public Library of Science Genetics* 4: e1000312. http://www.ncbi.nlm.nih.gov/pubmed/19096518.

Parham F, Austin C, Southall N, Huang R, Tice R, Portier C. (2009). Dose-response modeling of high-throughput screening data. *Journal of Biomolecular Screening* 14: 1216-1227. http://www.ncbi.nlm.nih.gov/pubmed/19828774.

Paris M, Laudet V. (2008). The history of a developmental stage: Metamorphosis in chordates. *Genesis* 46: 657-672. http://www.ncbi.nlm.nih.gov/pubmed/18932261.

Parng C, Seng WL, Semino C, McGrath P. (2002). Zebrafish: A preclinical model for drug screening. *Assay & Drug Development Technologies* 1: 41-48. http://www.ncbi.nlm.nih.gov/pubmed/15090155.

Patel CJ, Cullen MR. (2012). Genetic variability in molecular responses to chemical exposure. *Experientia* 101: 437-457. http://www.ncbi.nlm.nih.gov/pubmed/22945578.

Patel CJ, Chen R, Butte AJ. (2012a). Data-driven integration of epidemiological and toxicological data to select candidate interacting genes and environmental factors in association with disease. *Bioinformatics* 28: 1121-126. http://www.ncbi.nlm.nih.gov/pubmed/22689751.

Patel CJ, Cullen MR, Loannidis JP, Butte AJ. (2012b). Systematic evaluation of environmental factors: Persistent pollutants and nutrients correlated with serum lipid levels. *International Journal of Epidemiology* 41: 828-843. http://dx.doi.org/10.1093/ije/dys003.

Patel CJ, Chen R, Kodama K, Loannidis JP, Butte AJ. (2013). Systematic identification of interaction effects between genome-and environment-wide associations in type 2 diabetes mellitus. *Human genetics.*

Peden DB, Boehlecke B, Horstman D, Devlin R. (1997). Prolonged acute exposure to 0.16 ppm ozone induces eosinophilic airway inflammation in asthmatic subjects with allergies. *Journal of Allergy and Clinical Immunology* 100: 802-808. http://www.ncbi.nlm.nih.gov/pubmed/9438490.

Pedersen-Bjergaard J, Andersen MK, Andersen MT, Christiansen DH. (2008). Genetics of therapy-related myelodysplasia and acute myeloid leukemia. *Leukemia* 22: 240-248. http://www.ncbi.nlm.nih.gov/pubmed/18200041.

Perkins EJ, Ankley GT, Crofton KM, Garcia-Reyero N, Lalone CA, Johnson MS, et al. (2013). Current perspectives on the use of alternative species in human health and ecological risk assessments. *Environmental Health Perspectives.* http://www.ncbi.nlm.nih.gov/pubmed/23771518.

Physiome Project. (2013) Retrieved from http://physiomeproject.org/ (accessed March 28, 2013).

Rabinowitz JR, Goldsmith MR, Little SB, Pasquinelli MA. (2008). Computational molecular modeling for evaluating the toxicity of environmental chemicals: Prioritizing bioassay requirements. *Environmental Health Perspectives* 116: 573-577. http://www.ncbi.nlm.nih.gov/pubmed/18470285.

Rakyan VK, Down TA, Balding DJ, Beck S. (2011). Epigenome-wide association studies for common human diseases. *Nature Reviews Genetics* 12: 529-541. http://www.ncbi.nlm.nih.gov/pubmed/21747404.

Raldua D, Thienpont B, Babin PJ. (2012). Zebrafish eleutheroembryos as an alternative system for screening chemicals disrupting the mammalian thyroid gland morphogenesis and function. *Reproductive Toxicology* 33: 188-197. http://www.ncbi.nlm.nih.gov/pubmed/21978863.

Ramasamy A, Mondry A, Holmes CC, Altman DG. (2008). Key issues in conducting a meta-analysis of gene expression microarray datasets. *Public Library of Science Medicine* 5: e184. http://www.ncbi.nlm.nih.gov/pubmed/18767902.

Reaume CJ, Sokolowski MB. (2011). Conservation of gene function in behaviour. *Philosophical Transactions of the Royal Society London B: Biological Science* 366: 2100-2110. http://www.ncbi.nlm.nih.gov/pubmed/21690128.

Reif DM, Martin MT, Tan SW, Houck KA, Judson RS, Richard AM, et al. (2010). Endocrine profiling and prioritization of environmental chemicals using ToxCast data. *Environmental Health Perspectives* 118: 1714-1720. http://www.ncbi.nlm.nih.gov/pubmed/20826373.

Reuschenbach P, Silvani M, Dammann M, Warnecke D, Knacker T. (2008). ECOSAR model performance with a large test set of industrial chemicals. *Chemosphere* 71: 1986-1995. http://www.ncbi.nlm.nih.gov/pubmed/18262586.

Rosenbaum RK, Bachmann TM, Swirsky Gold L, Huijbregts MAJ, Jolliet O, Juraske R. (2008). USEtox -- the UNEP-SETAC toxicity model: Recommended characterization factors for human toxicity and freshwater ecotoxicicty in life cycle impact assessment. *International Journal of Life Cycle Assessment* 13: 532-546.

Rotroff DM, Wetmore BA, Dix DJ, Ferguson SS, Clewell HJ, Houck KA, et al. (2010). Incorporating human dosimetry and exposure into high-throughput in vitro toxicity screening. *Toxicological Sciences* 117: 348-358. http://www.ncbi.nlm.nih.gov/pubmed/20639261.

Rudel RA, Dodson RW, E. N, A.R. Z, J.G. B. (2008). Correlations between urinary phthalate metabolites and phthalates, estrogenic compounds 4-butyl phenol and o-phenyl phenol, and some pesticides in home indoor air and house dust. *Epidemiology* 19: S332.

Rung J, Cauchi S, Albrechtsen A, Shen L, Rocheleau G, Cavalcanti-Proenca C, et al. (2009). Genetic variant near IRS1 is associated with type 2 diabetes, insulin resistance and hyperinsulinemia. *Nature Genetics* 41: 1110-1115. http://www.ncbi.nlm.nih.gov/pubmed/19734900.

Rusyn I, Gatti DM, Wiltshire T, Kleeberger SR, Threadgill DW. (2010). Toxicogenetics: Population-based testing of drug and chemical safety in mouse models. *Pharmacogenomics* 11: 1127-1136. http://www.ncbi.nlm.nih.gov/pubmed/20704464.

Rusyn I, Sedykh A, Low Y, Guyton KZ, Tropsha A. (2012). Predictive modeling of chemical hazard by integrating numerical descriptors of chemical structures and short-term toxicity assay data. *Toxicological Sciences* 127: 1-9. http://dx.doi.org/10.1093/toxsci/kfs095.

SAB (Science Advisory Board). (2013). *Draft SAB advice on advancing the application of computational toxicology research for human health risk assessment.* Washington, DC. Retrieved from http://yosemite.epa.gov/sab/sabproduct.nsf/46963ceebabd621905256cae0053d5c6/F3315D0EE2EDC11285257B0A005F5B7E/$File/CompTox-edited+1-29-13.pdf (accessed February 26, 2013).

---

*This document is a draft for review purposes only and does not constitute Agency policy. Do not cite or quote.*

September 2013                                          106

Sagredo C, Ovrebo S, Haugen A, Fujii-Kuriyama Y, Baera R, Botnen IV, et al. (2006). Quantitative analysis of benzo[a]pyrene biotransformation and adduct formation in Ahr knockout mice. *Toxicology Letters* 167: 173-182. http://www.ncbi.nlm.nih.gov/pubmed/17049425.

Samuels MH, Luther M, Henry P, Ridgway EC. (1994). Effects of hydrocortisone on pulsatile pituitary glycoprotein secretion. *Journal of Clinical Endocrinology & Metabolism* 78: 211-215. http://www.ncbi.nlm.nih.gov/pubmed/8288706.

Sand S, Portier CJ, Krewski D. (2011). A signal-to-noise crossover dose as the point of departure for health risk assessment. *Environmental Health Perspectives* 119: 1766-1774. http://www.ncbi.nlm.nih.gov/pubmed/21813365.

Sarapura VD, Samuels MH, Ridgway EC. (2002). Thyroid-Stimulating Hormone. In S Melmed (Ed.), *The Pituatry* (Second ed., pp. 187-229). Malden, MA: Blackwell Science.

Schnatter AR, Glass DC, Tang G, Irons RD, Rushton L. (2012). Myelodysplastic syndrome and benzene exposure among petroleum workers: an international pooled analysis. *Journal of the National Cancer Institute* 104: 1724-1737. http://www.ncbi.nlm.nih.gov/pubmed/23111193.

Schreiber SL. (2003). The small-molecule approach to biology: Chemical genetics and diversity-oriented organic synthesis make possible the systematic exploration of biology. *Chemical & Engineering News* 81: 51-61.

Schug TT, Janesick A, Blumberg B, Heindel JJ. (2011). Endocrine disrupting chemicals and disease susceptibility. *Journal of Steroid Biochemistry and Molecular Biology* 127: 204-215. http://www.ncbi.nlm.nih.gov/pubmed/21899826.

Scott LJ, Mohlke KL, Bonnycastle LL, Willer CJ, Li Y, Duren WL, et al. (2007). A genome-wide association study of type 2 diabetes in Finns detects multiple susceptibility variants. *Science* 316: 1341-1345. http://www.ncbi.nlm.nih.gov/pubmed/17463248.

Sedykh A, Zhu H, Tang H, Zhang L, Richard A, Rusyn I, et al. (2011). Use of in vitro HTS-derived concentration-response data as biological descriptors improves the accuracy of QSAR models of in vivo toxicity. *Environmental Health Perspectives* 119: 364-370. http://www.ncbi.nlm.nih.gov/pubmed/20980217.

Selgrade MK, Cooper KD, Devlin RB, van Loveren H, Biagini RE, Luster MI. (1995). Immunotoxicity--bridging the gap between animal research and human health effects. *Fundamental and Applied Toxicology* 24: 13-21. http://www.ncbi.nlm.nih.gov/pubmed/7713335.

Serafimova R, Todorov M, Nedelcheva D, Pavlov T, Akahori Y, Nakai M, et al. (2007). QSAR and mechanistic interpretation of estrogen receptor binding. *SAR and QSAR in Environmental Research* 18: 389-421. http://www.ncbi.nlm.nih.gov/pubmed/17514577.

Shaffer CL, Scialis RJ, Rong H, Obach RS. (2012). Using Simcyp to project human oral pharmacokinetic variability in early drug research to mitigate mechanism-based adverse events. *Biopharmaceutics & Drug Disposition* 33: 72-84. http://www.ncbi.nlm.nih.gov/pubmed/22213407.

Shah et al. (submitted). OnToP: An ontology for toxicity pathways.

Shah I, Wambaugh J. (2010). Virtual tissues in toxicology. *Journal of Toxicology and Environmental Health Part B: Critical Reviews* 13: 314-328. http://www.ncbi.nlm.nih.gov/pubmed/20574905.

Sheldon LS, Cohen Hubal EA. (2009). Exposure as part of a systems approach for assessing risk. *Environmental Health Perspectives* 117: 119-1194. http://www.ncbi.nlm.nih.gov/pubmed/19672394.

Shen M, Zhang L, Lee KM, Vermeulen R, Hosgood HD, Li G, et al. (2011). Polymorphisms in genes involved in innate immunity and susceptibility to benzene-induced hematotoxicity. *Experimental & Molecular Medicine* 43: 374-378. http://www.ncbi.nlm.nih.gov/pubmed/21540635.

Shi L, Jones WD, Jensen RV, Harris SC, Perkins RG, Goodsaid FM, et al. (2008). The balance of reproducibility, sensitivity, and specificity of lists of differentially expressed genes in microarray studies. *BMC Bioinformatics* 9 Suppl 9: S10. http://www.ncbi.nlm.nih.gov/pubmed/18793455.

Sille FC, Thomas R, Smith MT, Conde L, Skibola CF. (2012). Post-GWAS functional characterization of susceptibility variants for chronic lymphocytic leukemia. *Public Library of Science One* 7: e29632. http://www.ncbi.nlm.nih.gov/pubmed/22235315.

Sipes NS, Martin MT, Kothiya P, Reif DM, Judson RS, Richard AM, et al. (2013). Profiling 976 ToxCast chemicals across 331 enzymatic and receptor signaling assays. *Chemical Research in Toxicology* 26: 878-895. http://www.ncbi.nlm.nih.gov/pubmed/23611293.

---

*This document is a draft for review purposes only and does not constitute Agency policy. Do not cite or quote.*

September 2013      107

Sipes NS, Padilla S, Knudsen TB. (2011). Zebrafish: As an integrative model for twenty-first century toxicity testing. *Birth Defects Research: Part C, Embryo Today* 93: 256-267. http://www.ncbi.nlm.nih.gov/pubmed/21932434.

Sistonen J, Sajantila A, Lao O, Corander J, Barbujani G, Fuselli S. (2007). CYP2D6 worldwide genetic variation shows high frequency of altered activity variants and no continental structure. *Pharmacogenetics and Genomics* 17: 93-101. http://www.ncbi.nlm.nih.gov/pubmed/17301689.

Skolness SY, Blanksma CA, Cavallin JE, Churchill JJ, Durhan EJ, Jensen KM, et al. (2013). Propiconazole inhibits steroidogenesis and reproduction in the fathead minnow (Pimephales promelas). *Toxicological Sciences.* http://www.ncbi.nlm.nih.gov/pubmed/23339182.

Sladek R, Rocheleau G, Rung J, Dina C, Shen L, Serre D, et al. (2007). A genome-wide association study identifies novel risk loci for type 2 diabetes. *Nature* 445: 881-885. http://www.ncbi.nlm.nih.gov/pubmed/17293876.

Smith MT, Zhang L, McHale CM, Skibola CF, Rappaport SM. (2011). Benzene, the exposome and future investigations of leukemia etiology. *Chemico-Biological Interactions* 192: 155-159. http://www.ncbi.nlm.nih.gov/pubmed/21333640.

Smith MV, Boyd WA, Kissling GE, Rice JR, Snyder DW, Portier CJ, et al. (2009). A discrete time model for the analysis of medium-throughput C. elegans growth data. *Public Library of Science One* 4: e7018. http://www.ncbi.nlm.nih.gov/pubmed/19753303.

Steinthorsdottir V, Thorleifsson G, Reynisdottir I, Benediktsson R, Jonsdottir T, Walters GB, et al. (2007). A variant in CDKAL1 influences insulin response and risk of type 2 diabetes. *Nature Genetics* 39: 770-775. http://www.ncbi.nlm.nih.gov/pubmed/17460697.

Takeuchi F, Serizawa M, Yamamoto K, Fujisawa T, Nakashima E, Ohnaka K, et al. (2009). Confirmation of multiple risk Loci and genetic impacts by a genome-wide association study of type 2 diabetes in the Japanese population. *Diabetes* 58: 1690-1699. http://www.ncbi.nlm.nih.gov/pubmed/19401414.

Taylor KW, Novak RF, Anderson HA, Birnbaum LS, Blystone C, Devito M, et al. (2013). Evaluation of the Association between Persistent Organic Pollutants (POPs) and Diabetes in Epidemiological Studies: A National Toxicology Program Workshop Review. *Environmental Health Perspectives* 121: 774-783. http://www.ncbi.nlm.nih.gov/pubmed/23651634.

Teschendorff AE, Widschwendter M. (2012). Differential variability improves the identification of cancer risk markers in DNA methylation studies profiling precursor cancer lesions. *Bioinformatics* 28: 1487-1494. http://www.ncbi.nlm.nih.gov/pubmed/22492641.

Thayer KA, Heindel JJ, Bucher JR, Gallo MA. (2012). Role of environmental chemicals in diabetes and obesity: A National Toxicology Program Workshop report. *Environmental Health Perspectives* 120: 779-789. http://dx.doi.org/10.1289/ehp.1104597.

Thienpont B, Tingaud-Sequeira A, Prats E, Barata C, Babin PJ, Raldua D. (2011). Zebrafish eleutheroembryos provide a suitable vertebrate model for screening chemicals that impair thyroid hormone synthesis. *Environmental Science and Technology* 45: 7525-7532. http://www.ncbi.nlm.nih.gov/pubmed/21800831.

Thomas D. (2010). Gene--environment-wide association studies: Emerging approaches. *Nature Reviews Genetics* 11: 259-272. http://www.ncbi.nlm.nih.gov/pubmed/20212493.

Thomas R, Phuong J, McHale CM, Zhang L. (2012). Using bioinformatic approaches to identify pathways targeted by human leukemogens. *International Journal of Environmental Research and Public Health* 9: 2479-2503. http://www.ncbi.nlm.nih.gov/pubmed/22851955.

Thomas RS, Allen BC, Nong A, Yang L, Bermudez E, Clewell HJ, III, et al. (2007). A method to integrate benchmark dose estimates with genomic data to assess the functional effects of chemical exposure. *Toxicological Sciences* 98: 240-248. http://dx.doi.org/10.1093/toxsci/kfm092.

Thomas RS, Bao W, Chu TM, Bessarabova M, Nikolskaya T, Nikolsky Y, et al. (2009). Use of short-term transcriptional profiles to assess the long-term cancer-related safety of environmental and industrial chemicals. *Toxicological Sciences* 112: 311-321. http://dx.doi.org/10.1093/toxsci/kfp233.

Thomas RS, Clewell HJ, 3rd, Allen BC, Yang L, Healy E, Andersen ME. (2012). Integrating pathway-based transcriptomic data into quantitative chemical risk assessment: A five chemical case study. *Mutation Research* 746: 135-143. http://www.ncbi.nlm.nih.gov/pubmed/22305970.

*This document is a draft for review purposes only and does not constitute Agency policy. Do not cite or quote.*

September 2013                                                                 108

Thomas RS, Clewell HJ, Allen BC, Wesselkamper SC, Wang NC, Lambert JC, et al. (2011). Application of transcriptional benchmark dose values in quantitative cancer and noncancer risk assessment. *Toxicological Sciences* 120: 194-205. http://dx.doi.org/10.1093/toxsci/kfq355.

Thomas RS, Philbert MA, Auerbach SS, Wetmore BA, Devito MJ, Cote I, et al. (2013a). Incorporating new technologies into toxicity testing and risk assessment: Moving from 21st century vision to a data-driven framework. *Toxicological Sciences.* http://www.ncbi.nlm.nih.gov/pubmed/23958734.

Thomas RS, Wesselkamper SC, Wang NC, Zhao QJ, Petersen DD, Lambert JC, et al. (2013b). Temporal concordance between apical and transcriptional points of departure for chemical risk assessment. *Toxicological Sciences* 134: 180-194. http://www.ncbi.nlm.nih.gov/pubmed/23596260.

Tice RR, Austin CP, Kavlock RJ, Bucher JR. (2013). Improving the human hazard characterization of chemicals: a tox21 update. *Environmental Health Perspectives* 121: 756-765. http://www.ncbi.nlm.nih.gov/pubmed/23603828.

Tietge JE, Degitz SJ, Haselman JT, Butterworth BC, Korte JJ, Kosian PA, et al. (2013). Inhibition of the thyroid hormone pathway in Xenopus laevis by 2-mercaptobenzothiazole. *Aquatic Toxicology* 126: 128-136. http://www.ncbi.nlm.nih.gov/pubmed/23178179.

Timpson NJ, Lindgren CM, Weedon MN, Randall J, Ouwehand WH, Strachan DP, et al. (2009). Adiposity-related heterogeneity in patterns of type 2 diabetes susceptibility observed in genome-wide association data. *Diabetes* 58: 505-510. http://www.ncbi.nlm.nih.gov/pubmed/19056611.

Tokar EJ, Diwan BA, Thomas DJ, Waalkes MP. (2012). Tumors and proliferative lesions in adult offspring after maternal exposure to methylarsonous acid during gestation in CD1 mice. *Archives of Toxicology* 86: 975-982. http://www.ncbi.nlm.nih.gov/pubmed/22398986.

Tokar EJ, Qu W, Waalkes MP. (2011). Arsenic, stem cells, and the developmental basis of adult cancer. *Toxicological Sciences* 120 Suppl 1: S192-203. http://www.ncbi.nlm.nih.gov/pubmed/21071725.

Torkamani A, Topol EJ, Schork NJ. (2008). Pathway analysis of seven common diseases assessed by genome-wide association. *Genomics* 92: 265-272. http://www.ncbi.nlm.nih.gov/pubmed/18722519.

Uehara T, Minowa Y, Morikawa Y, Kondo C, Maruyama T, Kato I, et al. (2011). Prediction model of potential hepatocarcinogenicity of rat hepatocarcinogens using a large-scale toxicogenomics database. *Toxicology and Applied Pharmacology* 255: 297-306. http://dx.doi.org/10.1016/j.taap.2011.07.001.

van Leeuwen DM, Pedersen M, Knudsen LE, Bonassi S, Fenech M, Kleinjans JC, et al. (2011). Transcriptomic network analysis of micronuclei-related genes: A case study. *Mutagenesis* 26: 27-32. http://www.ncbi.nlm.nih.gov/pubmed/21164179.

Venkatapathy R, Moudgal CJ, Bruce RM. (2004). Assessment of the oral rat chronic lowest observed adverse effect level model in TOPKAT, a QSAR software package for toxicity prediction. *Journal of Chemical Information and Computer Sciences* 44: 1623-1629. http://www.ncbi.nlm.nih.gov/pubmed/15446819.

Venkatapathy R, Wang NC. (2013). Developmental toxicity prediction. *Methods in Molecular Biology* 930: 305-340. http://www.ncbi.nlm.nih.gov/pubmed/23086848.

Visscher H, Ross CJ, Rassekh SR, Sandor GS, Caron HN, van Dalen EC, et al. (2013). Validation of variants in SLC28A3 and UGT1A6 as genetic markers predictive of anthracycline-induced cardiotoxicity in children. *Pediatr Blood Cancer* 60: 1375-1381. http://www.ncbi.nlm.nih.gov/pubmed/23441093.

Vogelstein B, Lane D, Levine AJ. (2000). Surfing the p53 network. *Nature* 408: 307-310. http://www.ncbi.nlm.nih.gov/pubmed/11099028.

Waits ER, Nebert DW. (2011). Genetic architecture of susceptibility to PCB126-induced developmental cardiotoxicity in zebrafish. *Toxicological Sciences* 122: 466-475. http://www.ncbi.nlm.nih.gov/pubmed/21613231.

Walker JD, Carlsen L. (2002). QSARs for identifying and prioritizing substances with persistence and bioconcentration potential. *SAR and QSAR in Environmental Research* 13: 713-725. http://www.ncbi.nlm.nih.gov/pubmed/12570048.

Walker JD, Carlsen L, Hulzebos E, Simon-Hettich B. (2002). Global government applications of analogues, SARs and QSARs to predict aquatic toxicity, chemical or physical properties, environmental fate parameters and health effects of organic chemicals. *SAR and QSAR in Environmental Research* 13: 607-616. http://www.ncbi.nlm.nih.gov/pubmed/12479375.

Wambaugh J, Shah I. (2010). Simulating microdosimetry in a virtual hepatic lobule. *Public Library of Science Computational Biology* 6: e1000756. http://www.ncbi.nlm.nih.gov/pubmed/20421935.

---

Wang I, Zhang B, Yang X, Stepaniants S, Zhang C, Meng Q, et al. (2012). Systems analysis of eleven rodent disease models reveals an inflammatome signature and key drivers. *Molecular Systems Biology* 8: 594. http://www.ncbi.nlm.nih.gov/pubmed/22806142.

Wang N, Jay Zhao Q, Wesselkamper SC, Lambert JC, Petersen D, Hess-Wilson JK. (2012a). Application of computational toxicological approaches in human health risk assessment I. A tiered surrogate approach. *Regulatory Toxicology and Pharmacology* 63: 10-19. http://www.ncbi.nlm.nih.gov/pubmed/22369873.

Wang N, Rice GE, Teuschler LK, Colman J, Yang RS. (2012b). An in silico approach for evaluating a fraction-based, risk assessment method for total petroleum hydrocarbon mixtures. *Journal of Toxicology* 2012: 410143. http://www.ncbi.nlm.nih.gov/pubmed/22496687.

Wang N, Venkatapathy R, Bruce RM, Moudgal C. (2011). Development of quantitative structure-activity relationship (QSAR) models to predict the carcinogenic potency of chemicals. II. Using oral slope factor as a measure of carcinogenic potency. *Regulatory Toxicology and Pharmacology* 59: 215-226. http://www.ncbi.nlm.nih.gov/pubmed/20951756.

Wanjek C. (2013). Systems Biology as Defined by National Institute of Health (NIH). Retrieved from http://irp.nih.gov/catalyst/v19i6/systems-biology-as-defined-by-nih (accessed April 2, 2013).

Warner CM, Gust KA, Stanley JK, Habib T, Wilbanks MS, Garcia-Reyero N, et al. (2012). A systems toxicology approach to elucidate the mechanisms involved in RDX species-specific sensitivity. *Environmental Science and Technology* 46: 7790-7798. http://www.ncbi.nlm.nih.gov/pubmed/22697906.

Weiss JN, Karma A, MacLellan WR, Deng M, Rau CD, Rees CM, et al. (2012). "Good enough solutions" and the genetics of complex diseases. *Circulation Research* 111: 493-504. http://www.ncbi.nlm.nih.gov/pubmed/22859671.

Wetmore BA, Wambaugh JF, Ferguson SS, Li L, Clewell HJ, 3rd, Judson RS, et al. (2013). Relative impact of incorporating pharmacokinetics on predicting in vivo hazard and mode of action from high-throughput in vitro toxicity assays. *Toxicological Sciences* 132: 327-346. http://www.ncbi.nlm.nih.gov/pubmed/23358191.

Wetmore BA, Wambaugh JF, Ferguson SS, Sochaski MA, Rotroff DM, Freeman K, et al. (2012). Integration of dosimetry, exposure, and high-throughput screening data in chemical toxicity assessment. *Toxicological Sciences* 125: 157-174. http://www.ncbi.nlm.nih.gov/pubmed/21948869.

WHO (World Health Organization). (2012). *State of the Science of Endocrine Disrupting Chemicals*. Retrieved from http://apps.who.int/iris/bitstream/10665/78102/1/WHO_HSE_PHE_IHE_2013.1_eng.pdf (accessed March 4, 2013).

Wignall J, Muratov E, Fourches D, Sedykh A, Tropsha A, Woodruff TJ, et al. (2012). Modeling toxicity values using chemical structure, in vitro screening, and in vivo toxicity data *51st Annual Meetign of the Society of Toxicology* Poster Abstract # 299. http://www.toxicology.org/AI/Pub/Tox/2012Tox.pdf.

Wignall J, Muratov E, Fourches D, Tropsha A, Woodruff TJ, Zeise L, et al. (2013). Conditional toxicity value (CTV) predictor for generating toxicity values for data sparse chemicals. *52nd Annual Meeting of the Society of Toxicology* Poster Abstract #2454. http://www.toxicology.org/AI/PUB/Tox/2013Tox.pdf.

Williams LM, Oleksiak MF. (2011). Ecologically and evolutionarily important SNPs identified in natural populations. *Molecular Biology and Evolution* 28: 1817-1826. http://www.ncbi.nlm.nih.gov/pubmed/21220761.

Wooding SP, Watkins WS, Bamshad MJ, Dunn DM, Weiss RB, Jorde LB. (2002). DNA sequence variation in a 3.7-kb noncoding sequence 5' of the CYP1A2 gene: Implications for human population history and natural selection. *American Journal of Human Genetics* 71: 528-542. http://www.ncbi.nlm.nih.gov/pubmed/12181774.

Woodruff TJ, Sutton P. (2011). An evidence-based medicine methodology to bridge the gap between clinical and environmental health sciences. *Health Affairs (Millwood)* 30: 931-937. http://www.ncbi.nlm.nih.gov/pubmed/21555477.

Wu W, Doreswamy V, Diaz-Sanchez D, Samet JM, Kesic M, Dailey L, et al. (2011). GSTM1 modulation of IL-8 expression in human bronchial epithelial cells exposed to ozone. *Free Radical Biology and Medicine* 51: 522-529. http://www.ncbi.nlm.nih.gov/pubmed/21621609.

Yeung KY, Dombek KM, Lo K, Mittler JE, Zhu J, Schadt EE, et al. (2011). Construction of regulatory networks using expression time-series data of a genotyped population. *Proceedings of the National Academy of Sciences USA* 108: 19436-19441. http://www.ncbi.nlm.nih.gov/pubmed/22084118.

Zacharewski T, Teuschler L, Cote I, Burgoon L. (submitted). Improving cumulative risk assessment through systems and network biology driven data mining.

*This document is a draft for review purposes only and does not constitute Agency policy. Do not cite or quote.*

September 2013          110

Zeggini E, Weedon MN, Lindgren CM, Frayling TM, Elliott KS, Lango H, et al. (2007). Replication of genome-wide association signals in UK samples reveals risk loci for type 2 diabetes. *Science* 316: 1336-1341. http://www.ncbi.nlm.nih.gov/pubmed/17463249.

Zeise L, Bois FY, Chiu WA, Hattis D, Rusyn I, Guyton KZ. (2012). Addressing human variability in next-generation human health risk assessments of environmental chemicals. *Environmental Health Perspectives* Epub doi: 10.1289/ehp.1205687.

Zhang L, McHale CM, Rothman N, Li G, Ji Z, Vermeulen R, et al. (2010). Systems biology of human benzene exposure. *Chemico-Biological Interactions* 184: 86-93. http://www.ncbi.nlm.nih.gov/pubmed/20026094.

Zhang Q, Bhattacharya S, Andersen ME, Conolly RB. (2010). Computational systems biology and dose-response modeling in relation to new directions in toxicity testing. *Journal of Toxicology and Environmental Health Part B: Critical Reviews* 13: 253-276. http://www.ncbi.nlm.nih.gov/pubmed/20574901.

Zhu H, Rao RS, Zeng T, Chen L. (2012). Reconstructing dynamic gene regulatory networks from sample-based transcriptional data. *Nucleic Acids Research* 40: 10657-10667. http://www.ncbi.nlm.nih.gov/pubmed/23002138.

Zhuo W, Zhang L, Zhu B, Qiu Z, Chen Z. (2012). Association between CYP1A1 Ile462Val variation and acute leukemia risk: Meta-analyses including 2164 cases and 4160 controls. *Public Library of Science One* 7: e46974. http://www.ncbi.nlm.nih.gov/pubmed/23056546.

Zoeller RT, Crofton KM. (2005). Mode of action: Developmental thyroid hormone insufficiency--neurological abnormalities resulting from exposure to propylthiouracil. *Critical Reviews in Toxicology* 35: 771-781. http://www.ncbi.nlm.nih.gov/pubmed/16417044.

Zoeller RT, Dowling AL, Herzig CT, Iannacone EA, Gauger KJ, Bansal R. (2002). Thyroid hormone, brain development, and the environment. *Environmental Health Perspectives* 110 Suppl 3: 355-361. http://www.ncbi.nlm.nih.gov/pubmed/12060829.

Zoeller RT, Rovet J. (2004). Timing of thyroid hormone action in the developing brain: Clinical observations and experimental findings. *Journal of Neuroendocrinology* 16: 809-818. http://www.ncbi.nlm.nih.gov/pubmed/15500540.

[page intentionally left blank]

# Appendix A. Technical Papers Supporting the NexGen Report

| | Technical Papers Supporting the Report |
|---|---|
| **Preparation for Prototype Development** | *Advancing the Next Generation of Risk Assessment* by Ila Cote, Paul Anastas, Linda Birnbaum, Becki Clark, David Dix, Stephen Edwards, and Peter Preuss (2012) |
| | *Advancing the Next Generation (NexGen) of Risk Assessment: The Prototypes Workshop* by EPA (2010) |
| | *Summary Report of Advancing the Next Generation of Risk Assessment Public Dialogue Conference* by EPA (2011a) |
| | *A Framework for the Next Generation of Risk Assessment* by Daniel Krewski, Margit Westphal, Greg Paoli, Maxine Croteau, Mustafa Al-Zoughool, Mel Andersen, Weihsueh Chiu, Lyle Burgoon, and Ila Cote (2013) |
| | *Reconsideration of Important Risk Assessment Issues Informed by Molecular Systems Biology* by Daniel Krewski, Melvin Andersen, Kim Boekelheide, Frederic Bois, Lyle Burgoon, Weihsueh Chiu, Michael DeVito, Hisham El-Masri, Lynn Flowers, Michael Goldsmith, Derek Knight, Thomas Knudsen, William Lefew, Greg Paoli, Edward Perkins, Ivan Rusyn, Cecilia Tan, Linda Teuschler, Russell Thomas, Maurice Whelan, Timothy Zacharewski, Lauren Zeise, and Ila Cote (in preparation) |
| **Tier 3 Prototypes: Leukemia & Benzene, Lung Injury & Ozone, Liver Cancer & B[a]P/PAHs** | *Characterization of Changes in Gene Expression and Biochemical Pathways at Low Levels of Benzene Exposure* by Reuben Thomas, Alan Hubbard, Cliona McHale, Luoping Zhang, Stephen Rappaport, Qing Lan, Nathaniel Rothman, Kathryn Guyton, Jennifer Jinot, Babasaheb Sonawane, and Martyn Smith (in preparation) |
| | *Current Understanding of the Mechanism of Benzene-Induced Leukemia in Humans: Implications for Risk Assessment* by Cliona McHale, Luoping Zhang, and Martyn Smith (2012) |
| | *Benzene, the Exposome and Future Investigations of Leukemia Etiology* by Martyn Smith, Luoping Zhang, Cliona McHale, Christine Skibola, and Stephen Rappaport (2011) |
| | *Global Gene Expression Profiling of a Population Exposed to a Range of Benzene Levels* by Cliona McHale, Luoping Zhang, Qing Lan, Roel Vermeulen, Guilan Li, Alan Hubbard, Kristin Porter, Reuben Thomas, Christopher Portier, Min Shen, Stephen Rappaport, Songnian Yin, Martyn Smith, and Nathaniel Rothman (2011) |
| | *Temporal Profile of Gene Expression Alterations In Primary Human Bronchial Epithelial Cells Following In Vivo Exposure to 0.3 ppm Ozone* (meeting abstract) by Kelly Duncan, James Crooks, David Miller, Lyle Burgoon, Michael Schmitt, Stephen Edwards, David Diaz-Sanchez, and Robert Devlin (2013) |
| | *Transcriptional Profiling of Ozone-Induced Stress Responses in Primary Human Bronchial Epithelial Cells Cultured at an Air-Liquid Interface* by Kelly Duncan et al. (in preparation) |
| | *Ozone-induced Inflammation Is not Mediated via the Canonical NF-κB Pathway in Humans* by David Miller, Stephen Edwards, Lyle Burgoon, Rory Conolly, William Lefew, Kelly Duncan, Robert Devlin, and James Samet (in preparation) |
| | *Systems Biology Informed Assessment of Benzo[a]pyrene/Polycyclic Aromatic Hydrocarbons and Liver Cancer* by Lyle Burgoon and Emma McConnell (in preparation) |
| | *IRIS Toxicological Review of Benzo[a]pyrene* (Public Comment Draft). U.S. Environmental Protection Agency, Washington, DC, EPA/635/R-13/138a-b (2013). |

*This document is a draft for review purposes only and does not constitute Agency policy. Do not cite or quote.*

September 2013    A-1

| Technical Papers Supporting the Report | |
|---|---|
| **Tier 2 Prototypes Knowledge Mining Diabetes/Obesity Example** | *Data Mining Informed Risk Analysis of Environmental and Genetic Factors Associated with Type 2 Diabetes Mellitus* by Lyle Burgoon (in preparation) |
| | *Data Mining NHANES to Identify Environmental Chemical and Disease Associations* by Shannon Bell and Stephen Edwards (in preparation) |
| | *Systematic Identification of Interaction Effects Between Genome- and Environment-Wide Associations in Type 2 Diabetes Mellitus* by Chirag Patel, Rong Chen, Keiichi Kodama, John Ioannidis, and Atul Butte (2013) |
| | *Data-Driven Integration of Epidemiological and Toxicological Data to Select Candidate Interacting Genes and Environmental Factors in Association with Disease* by Chirag Patel, Rong Chen, and Atul Butte (2012a) |
| | *Genetic Variability in Molecular Responses to Chemical Exposure* by Chirag Patel and Mark Cullen (2012) |
| **Tier 2 Prototypes Short-term in Vivo Nonmammalian** | *Role of Environmental Chemicals in Diabetes and Obesity: An NTP Workshop Review* by Kristina Thayer, Jerrold Heindel, John Bucher, and Michael Gallo (2012) |
| | *Current Perspectives on the Use of Alternative Species in Human Health and Ecological Hazard Assessments* by Edward Perkins, Gerald Ankley, Kevin Crofton, Natàlia Garcia-Reyero, Carlie LaLone, Mark Johnson, Joseph Tietge, and Daniel Villeneuve (2013) |
| | *Propiconazole Inhibits Steroidogenesis and Reproduction in the Fathead Minnow (Pimephales promelas)* by Sarah Skolness, Chad Blanksma, Jenna Cavallin, Jessica Churchill, Elizabeth Durhan, Kathleen Jensen, Rodney Johnson, Michael Kahl, Elizabeth Makynen, Daniel Villeneuve, and Gerald Ankley (2013) |
| | *Zebrafish Developmental Screening of the ToxCast™ Phase I Chemical Library* by Stephanie Padilla, Daniel Corum, Beth Padnos, Deborah Hunter, Andrew Beam, Keith Houck, Nisha Sipes, Nicole Kleinstreuer, Thomas Knudsen, David Dix, and David Reif (2012) |
| | *A Systems Toxicology Approach to Elucidate the Mechanisms Involved in RDX Species-Specific Sensitivity* by Christopher Warner, Kurt Gust, Jacob Stanley, Tanwir Habib, Mitchell Wilbanks, Natàlia Garcia-Reyero, and Edward Perkins (2012) |
| **Tier 2 Prototypes Short-term In Vivo Mammalian** | *Development of a Paradigm for the Next Generation of Chemical Risk Assessment: Short-term In Vivo Models for Tier 2 Assessments* by Michael DeVito, Russell Thomas, and Jason Lambert (in preparation) |
| | *Incorporating New Technologies into Toxicity Testing and Risk Assessment: Moving from 21st Century Vision to a Data-Driven Framework* by Russell Thomas, Martin Philbert Scott Auerbach, Barbara Wetmore, Michael Devito, Ila Cote, Craig Rowlands, Maurice Whelan, Sean Hays, Melvin Andersen, Bette Meek, Lawrence Reiter, Jason Lambert, Harvey Clewell III, Martin Stephens, Jay Zhao, Scott Wesselkamper, Lynn Flowers, Edward Carney, Timothy Pastoora, Dan Petersen, Carole Yauk, and Andy Nong (2013a) |
| | *Temporal Concordance Between Apical and Transcriptional Points of Departure for Chemical Risk Assessment* by Russell Thomas, Scott Wesselkamper, Nina Wang, Jay Zhao, Dan Peterson, Jason Lambert, Ila Cote, Yang Longlong, Eric Healy, Michael Black, Harvey Clewell, Bruce Allen, and Melvin Andersen (2013b) |
| | *Integrating Pathway-Based Transcriptomic Data into Quantitative Chemical Risk Assessment: A Five Chemical Case Study* by Russell Thomas, Harvey Clewell III, Bruce Allen, Longlong Yang, Eric Healy, and Melvin Andersen (2012) |
| | *Application of Transcriptional Benchmark Dose Values in Quantitative Cancer and Noncancer Risk Assessment* by Russell Thomas, Harvey Clewell III, Bruce Allen, Scott Wesselkamper, Nina Ching Wang, Jason Lambert, Janet Hess-Wilson, Jay Zhao, and Melvin Andersen (2011) |

*This document is a draft for review purposes only and does not constitute Agency policy. Do not cite or quote.*

September 2013      A-2

## Technical Papers Supporting the Report

| | |
|---|---|
| **Tier 1 Prototypes**<br>*Integration of QSAR and Various Biological Data Streams* | *Predictive QSAR Modeling: Methods and Applications in Drug Discovery and Chemical Risk Assessment* by Alexander Golbraikh, Xiang Simon Wang, Hao Zhu, and Alexander Tropsha (2012) |
| | *Developmental Toxicity Prediction* by Raghuraman Venkatapathy and Nina Wang (2013) |
| | *Predictive Modeling of Chemical Hazard by Integrating Numerical Descriptors of Chemical Structures and Short-term Toxicity Assay Data* by Ivan Rusyn, Alexander Sedykh, Yen Low, KZ Guyton, and Alexander Tropsha (2012) |
| | *An In Silico Approach for Evaluating a Fraction-Based, Risk Assessment Method for Total Petroleum Hydrocarbon Mixtures* by Nina Ching Wang, Glenn Rice, Linda Teuschler, Joan Colman, and Raymond Yang (2012b) |
| | *Application of Computational Toxicological Approaches in Human Health Risk Assessment I. A Tiered Surrogate Approach* by Nina Ching Yi Wang, Jay Zhao, Scott Wesselkamper, Jason Lambert, Dan Petersen, and Janet Hess-Wilson (2012a) |
| | *Development of Quantitative Structure-Activity Relationship (QSAR) Models to Predict the Carcinogenic Potency of Chemicals. II. Using Oral Slope Factor as a Measure of Carcinogenic Potency* by Nina Ching Yi Wang, Raghuraman Venkatapathy, Robert Mark Bruce, and Chandrika Moudgal (2011) |
| **Tier 1 Prototypes**<br>*High-throughput Screening* | *Perspectives on Validation of High-Throughput Assays Supporting 21st Century Toxicity Testing* by Richard Judson, Robert Kavlock, Matthew Martin, David Reif, Keith Houck, Thomas Knudsen, Ann Richard, Raymond Tice, Maurice Whelan, Menghang Xia, Ruili Huang, Christopher Austin, George Daston, Thomas Hartung, John Fowle III, William Wooge, Weida Tong, and David Dix (2013) |
| | *Estimating Toxicity-Related Biological Pathway Altering Doses for High-Throughput Chemical Risk Assessment* by Richard Judson, Robert Kavlock, Woodrow Setzer, Elaine Cohen Hubal, Matthew Martin, Thomas Knudsen, Keith Houck, Russell Thomas, Barbara Wetmore, and David Dix (2011) |
| **Key Risk Assessment Issues** | *Addressing Human Variability in Next Generation Health Assessments of Environmental Chemicals* by Lauren Zeise, Frederic Bois, Weihsueh Chiu, Dale Hattis, Ivan Rusyn, and Kathryn Guyton (2012) |
| | *Quantitative High-Throughput Screening for Chemical Toxicity in a Population-Based In Vitro Model* by Eric Lock, Nour Abdo, Ruili Huang, Menghang Xia, Oksana Kosyk, Shannon O'Shea, Yi-Hui Zhou, Alexander Sedykh, Alexander Tropsha, Christopher Austin, Raymond Tice, Fred Wright, and Ivan Rusyn (2012) |
| | *Predicting Later-Life Outcomes of Early-Life Exposures* by Kim Boekelheide, Bruce Blumberg, Robert Chapin, Ila Cote, Joseph Graziano, Amanda Janesick, Robert Lane, Karen Lillycrop, Leslie Myatt, Christopher States, Kristina Thayer, Michael Waalkes, and John Rogers (2012) |
| | *In Vitro Screening for Population Variability in Chemical Toxicity* by Shannon O'Shea, John Schwarz, Oksana Kosyk, Pamela Ross, Min Jin Ha, Fred Wright, and Ivan Rusyn (2011) |
| | *Improving Cumulative Risk Assessment Through Systems and Network Biology Driven Data Mining* by Timothy Zacharewski, Ila Cote, Linda Teuschler, and Lyle Burgoon (submitted) |
| | *The Role of Advanced Biological Methods and Data in Regulatory Rationality* by Douglas Crawford-Brown (2013) |
| | *Incorporating New Technologies into Toxicity Testing and Risk Assessment: Moving from 21st Century Vision to a Data-Driven Framework* T by Russell S. Thomas, Martin Philbert, Scott Auerbach, Barbara Wetmore, Michael Devito, Ila Cote, et al. (2013) |

Note: EPA also thanks Christine Sofge, Paul Schulte, and Ainsley Weston for sharing their pre-publication draft manuscript.

# Appendix B. Glossary

| Glossary Term | Description |
|---|---|
| adverse outcome pathway (AOP) | An adverse outcome pathway is the mechanistic or predictive relationship between an initial chemical-biological interaction (i.e., molecular initiating event[s]) and subsequent perturbation to cellular functions sufficient to elicit disruptions at higher levels of organization, culminating in an adverse phenotypic outcome in an individual and population relevant to risk assessment (i.e., disease progression or organ dysfunction in humans).<br><br>Ankley GT; Bennett RS; Erickson RJ; Hoff DJ; Hornung MW; Johnson RD; Mount DR; Nichols JW; Russom CL; Schmieder PK; Serrrano JA; Tietge JE; Villeneuve DL. (2010). Adverse outcome pathways: A conceptual framework to support ecotoxicology research and risk assessment. Environmental Toxicology and Chemistry 29 (3): 730-741.<br><br>http://service004.hpc.ncsu.edu/toxicology/websites/journalclub/linked_files/Fall10/Environ%20Toxicol%20Chem%202010%20Ankley.pdf. |
| ArrayTrack™ | Publicly available toxicogenomics software for DNA microarrays. It contains three integrated components: (1) a database (MicroarrayDB) that stores microarray data and associated toxicological information; (2) tools (TOOL) for data visualization and analysis; and (3) libraries (LIB) that provide curated functional data from public databases for data interpretation. Using ArrayTrack™, an analysis method can be selected from TOOL and applied to selected microarray data stored in the MicroarrayDB. Analysis results can be linked directly to pathways, gene ontology, and other functional information stored in LIB.<br><br>Food and Drug Administration (FDA). (2012). ArrayTrack™ FAQs. Available online at<br>http://www.fda.gov/ScienceResearch/BioinformaticsTools/Arraytrack/ucm135070.htm (accessed September 27, 2012). |
| assay | 1. The process of quantitative or qualitative analysis of a component of a sample; or<br>2. Results of a quantitative or qualitative analysis of a component of a sample.<br><br>National Library of Medicine. (2012). IUPAC Glossary of Terms Used in Toxicology, 2nd Ed. Available online at http://sis.nlm.nih.gov/enviro/iupacglossary/frontmatter.html (accessed September 27, 2012). |
| Bayesian Network | A graph-based model of joint multivariate probability distributions that captures properties of conditional independence between variables.<br><br>Friedman N; Linial M; Nachman I; Pe'er D. (2000). Using Bayesian networks to analyze expression data. Journal of Computational Biology 7 (3-4): 601-620. |

| Glossary Term | Description |
|---|---|
| **Bayesian Network reconstruction** | The process of integrating Bayesian Network data using a software program to generate gene causal networks predictive of complex phenotypes. |
| | Wang I.; Zhang B; Yang X; Stepaniants S; Zhang C; Meng Q; Peters M; He Y; Ni C; Slipetz D; Crackower MA; Houshyar H; Tan CM; Asante-Appiah E; O'Neill G; Luo MJ; Theiringer R; Yuan J; Chiu C; Lum PY; Lamb J; Boie Y; Wilkinson HA; Schadt E; Dai H; Roberts C. (2012). Systems analysis of eleven rodent disease models reveals an inflammatome signature and key drivers. Molecular Systems Biology 8 594. |
| **bioinformatics** | A field of biology in which complex multivariable data from high-throughput screening and genomic assays are interpreted in relation to target identification and effects of sustained perturbations on organs and tissues to make biological discoveries or predictions. This field encompasses all computational methods and theories applicable to molecular biology and areas of computer-based techniques for solving biological problems, including manipulation of models and data sets. |
| | National Center for Biotechnology Information (NCBI). (2012). Bioinformatics. Available online at http://www.ncbi.nlm.nih.gov/mesh?term=bioinformatics (accessed September 27, 2012). |
| **biological assay (bioassay)** | A method of measuring the effects of a biologically active substance using an intermediate *in vivo* or *in vitro* tissue or cell model under controlled conditions. It includes virulence studies in animal fetuses *in utero*, mouse convulsion bioassay of insulin, quantitation of tumor-initiator systems in mouse skin, calculation of potentiating effects of a hormonal factor in an isolated strip of contracting stomach muscle, etc. |
| | National Center for Biotechnology Information (NCBI). (2012). Biological Assay. Available online at http://www.ncbi.nlm.nih.gov/mesh?term=bioassay (accessed September 27, 2012). |
| **biological pathway altering dose (BPAD)** | The provisional acceptable exposure level at the low end of the distribution of the external dose required to perturb a biological pathway, accounting for uncertainty and variability. |
| | Judson RS; Kavlock RJ; Setzer RW; Hubal EA; Martin MT; Knudsen TB; Houck KA; Thomas RS; Wetmore BA; Dix DJ. (2011). Estimating toxicity-related biological pathway altering doses for high-throughput chemical risk assessment. Chem Res Toxicol 24 (4): 451-462. http://dx.doi.org/10.1021/tx100428e |
| **biomarkers** | Measurable and quantifiable biological parameters (e.g., specific enzyme concentrations, specific hormone concentrations, a specific gene phenotype distribution in a population, presence of biological substances) that serve as indices for health- and physiology-related assessments, such as disease risk, psychiatric disorders, environmental exposure and its effects, disease diagnosis, metabolic processes, substance abuse, pregnancy, cell line development, epidemiologic studies. |
| | National Center for Biotechnology Information (NCBI). (2012). Biological Markers. Available online at http://www.ncbi.nlm.nih.gov/mesh?term=biological%20markers (accessed September 27, 2012). |

*This document is a draft for review purposes only and does not constitute Agency policy. Do not cite or quote.*

September 2013        B-2

| Glossary Term | Description |
|---|---|
| cell biology | The study of the structure, behavior, growth, reproduction, and pathology of cells; and the function and chemistry of cellular components. |
| | National Center for Biotechnology Information (NCBI). (2012). Cell Biology Available online at http://www.ncbi.nlm.nih.gov/mesh?term=cell%20biology (accessed September 27, 2012). |
| Chemical Effects in Biological Systems (CEBS) database | An NIH/NIEHS publicly available toxicogenomic database that houses data of interest to environmental health scientists. CEBS has received depositions of data from academic, industrial, and governmental laboratories. CEBS is designed to display data in the context of biology and study design, and to permit data integration across studies for novel meta-analysis. |
| | National Institute for Environmental Health Sciences (NIEHS). (2012). Chemical Effects in Biological Systems (CEBS). Available online at http://www.niehs.nih.gov/research/resources/databases/cebs/index.cfm (accessed September 27, 2012). |
| Comparative Toxicogenomic Database (CTD)™ | A publicly available toxicogenomic database on the National Library of Medicine's (NLM) Toxicology Data Network (TOXNET®). The CTD™ elucidates molecular mechanisms by which environmental chemicals affect human disease. It contains manually curated data describing cross-species chemical-gene/protein interactions and chemical- and gene-disease relationships. The results provide insight into the molecular mechanisms underlying variable susceptibility and environmentally influenced diseases. These data also will provide insights into complex chemical-gene and protein interaction networks. |
| | National Library of Medicine (NLM). (2012). Fact Sheet. Comparative Toxicogenomics Database (CTD)™. Available online at http://www.nlm.nih.gov/pubs/factsheets/ctdfs.html (accessed September 27, 2012). |
| computational models | Computerized predictive tools. Sometimes referred to as "in silico" models. |
| | U.S. Environmental Protection Agency (EPA). (2012). Glossary of Terms: Methods of Toxicity Testing and Risk Assessment. Available online at http://www.epa.gov/opp00001/science/comptox-glossary.html (accessed April 2, 2013). |

*This document is a draft for review purposes only and does not constitute Agency policy. Do not cite or quote.*

September 2013                                                                 B-3

| Glossary Term | Description |
|---|---|
| decision context | Decision context seeks to understand and describe what management decisions are being made, why these decisions are made, and the relationship of these decisions to previous and anticipated decisions. For example, decision context tries to answer some of the following questions: Are risks being ranked; if so, why? How will risk information be used in future decisions? Is a change in policy or management under consideration; and if so, what is driving the change and what are the underlying policy objectives? What is the general scope of alternatives under consideration and why?<br><br>Decision context defines the roles and responsibilities of the ultimate decision maker, stakeholders, and key technical experts in relation to the decision process. Decision context also identifies the constraints within which a decision must be made and outputs that will result from the decision.<br><br>Structured Decision Making (SDM). (2008). Steps in the Decision Process: Introduction. Available online at http://www.structureddecisionmaking.org/DecisionContext.htm (accessed March 19, 2013). |
| DNA microarray | A grid of nucleic acid molecules of known sequence linked to a solid substrate, which can be probed with a sample containing either messenger RNA or complementary DNA from a cell or tissue to reveal changes in gene expression relative to a control sample. Microarray technology, also known as "DNA gene chip" technology, enables the expression of many thousands of genes to be assessed in a single experiment. DNA microarrays exploit the ability of complementary strands of nucleic acids to base-pair with each other and bind. For example, ATATGCGC will bind to its complement (TATACGCG) with a certain affinity. DNA copies (cDNAs) are melted, or denatured, to single strands, which then can be used to bind to, or hybridize with, fluorescently labeled nucleic acid samples from cancerous or normal cells. After washing away the unbound molecules, bound fluorescent nucleic acid samples can be identified by laser microscopy. Fluorescent dots indicate expressed genes, and differences in microarray patterns between normal and cancerous cells can be quickly identified.<br><br>National Library of Medicine. (2012). IUPAC Glossary of Terms Used in Toxicology, 2nd Ed. Available online at http://sis.nlm.nih.gov/enviro/iupacglossary/frontmatter.html (accessed September 28, 2012). |
| Enzyme-Linked Immunosorbent Assay (ELISA) | An immunoassay utilizing an antibody labeled with an enzyme marker such as horseradish peroxidase. Although either the enzyme or the antibody is bound to an immunosorbent substrate, they both retain their biologic activity; the change in enzyme activity as a result of the enzyme-antibody-antigen reaction is proportional to the concentration of the antigen and can be measured spectrophotometrically or with the naked eye. Many variations of the method have been developed.<br><br>National Center for Biotechnology Information (NCBI). (2012). Enzyme-Linked Immunosorbent Assay. Available online at http://www.ncbi.nlm.nih.gov/mesh?term=elisa (accessed September 27, 2012). |

*This document is a draft for review purposes only and does not constitute Agency policy. Do not cite or quote.*

September 2013                                                                                              B-4

| Glossary Term | Description |
|---|---|
| epigenetics | An emerging field of science that studies heritable changes caused by the activation and deactivation of genes with no change in the underlying DNA sequence of the organism. The word is Greek in origin and literally means over and above (epi) the genome.<br><br>National Human Genome Research Institute (NHGRI). (2012). Talking Glossary of Genetic Terms. Available online at http://www.genome.gov/glossary/index.cfm?id=528&textonly=true (accessed September 27, 2012). |
| functional genomics | The study of dynamic cellular processes such as gene transcription, translation, and gene product interactions that define an organism.<br><br>The National Institutes of Health (NIH). (2009). Genomics and Advanced Technologies. Available online at http://www.niaid.nih.gov/topics/pathogengenomics/Pages/definitions.aspx (accessed September 28, 2012). |
| gene-environment interaction | The combined effects of genotypes and environmental factors on phenotypic characteristics.<br><br>National Center for Biotechnology Information (NCBI). (2012). Gene-Environment Interaction. Available online at http://www.ncbi.nlm.nih.gov/mesh?term=gene%20environment%20interaction (accessed September 28, 2012). |
| gene expression | The phenotypic manifestation of a gene or genes by the processes of genetic transcription and genetic translation.<br><br>National Center for Biotechnology Information (NCBI). (2012). Gene Expression. Available online at http://www.ncbi.nlm.nih.gov/mesh/68015870 (accessed September 28, 2012). |
| Gene Expression Omnibus (GEO) | A public repository that archives and freely distributes microarray, next-generation sequencing, and other forms of high-throughput functional genomic data submitted by the scientific community. In addition to data storage, a collection of Web-based interfaces and applications is available to help users query and download the studies and gene expression patterns stored in GEO.<br><br>National Center for Biotechnology Information (NCBI). (2012). Gene Expression Omnibus. Frequently Asked Questions. Available online at http://www.ncbi.nlm.nih.gov/geo/info/faq.html (accessed September 27, 2012). |

*This document is a draft for review purposes only and does not constitute Agency policy. Do not cite or quote.*

September 2013        D-5

| Glossary Term | Description |
|---|---|
| **Gene Ontology (GO) database** | A product of the Gene Ontology (GO) project. The GO project provides structured, controlled vocabularies and classifications that cover several domains of molecular and cellular biology and are freely available for community use in the annotation of genes, gene products, and sequences. Many model organism databases and genome annotation groups use the GO database and contribute their annotation sets to the GO resource. The GO database integrates the vocabularies and contributed annotations and provides full access to this information in several formats. Members of the GO Consortium continuously work collectively, involving outside experts as needed, to expand and update the GO vocabularies. The GO Web resource also provides access to extensive documentation about the GO project and links to applications that use GO data for functional analyses.

Gene Ontology Consortium. (2004). The Gene Ontology (GO) database and informatics resource. Nucleic Acids Research 32: Database issue D258-261. |
| **genetics** | The branch of science concerned with the means and consequences of transmission and generation of the components of biological inheritance. Used for mechanisms of heredity and the genetics of organisms, for the genetic basis of normal and pathologic states, and for the genetic aspects of endogenous chemicals. It includes biochemical and molecular influence on genetic material.

National Center for Biotechnology Information (NCBI). (2012). Genetics. Available online at http://www.ncbi.nlm.nih.gov/mesh?term=genetics (accessed September 27, 2012). |
| **genome-wide association study (GWAS)** | An approach used in genetics research to associate specific genetic variations with particular diseases. The method involves scanning the genomes from many different people and looking for genetic markers that can be used to predict the presence of a disease. Once such genetic markers are identified, they can be used to understand how genes contribute to the disease and develop better prevention and treatment strategies.

National Institutes of Health (NIH). (2012). Talking Glossary of Genetic Terms: Genome-wide Association Studies (GWAS). National Human Genome Research Institute. Available online at http://www.genome.gov/glossary/index.cfm?id=91&textonly=true (accessed September 27, 2012). |
| **green chemistry** | The design of chemical products and processes to reduce or eliminate the use and generation of hazardous substances. Green Chemistry framework includes three main principles: (1) to incorporate sustainable designs across all stages of the chemical lifecycle, (2) to reduce the hazard of chemical products and processes by design, and (3) to work as a cohesive set of design criteria. Twelve design criteria have been developed to fulfill these three principles (prevention, atom economy, less hazardous chemical synthesis, designing safer chemicals, safer solvents and auxiliaries, design for energy efficiency, use of renewable feedstocks, reduce derivatives, catalysis, design for degradation, real-time analysis for pollution prevention, and inherently safer chemistry for accident prevention).

Anastas, P, Eghbali, N. (2010). Green chemistry: Principles and practice. Chem Soc Rev 39 (1): 301-312. |

| Glossary Term | Description |
|---|---|
| high-throughput screening (HTS) | A rapid method of measuring the effect of an agent in a biological or chemical assay. The assay usually involves some form of automation or a way to conduct multiple assays at the same time using sample arrays. |
| | National Center for Biotechnology Information (NCBI). (2012). High-Throughput Screening Assays. Available online at http://www.ncbi.nlm.nih.gov/mesh?term=high%20throughput%20screening%20method (accessed September 27, 2012). |
| in silico | Referring to or describing data generated and analyzed using computer modeling and information technology. |
| | National Library of Medicine. (2012). IUPAC Glossary of Terms Used in Toxicology, 2nd Ed. Available online at http://sis.nlm.nih.gov/enviro/iupacglossary/frontmatter.html (accessed September 27, 2012). |
| IVIV extrapolation (IVIVE) | A method that uses determinations of protein binding, liver/kidney clearance, and oral uptake to estimate ranges of oral human exposures leading to tissue/plasma concentrations similar to *in vitro* point-of-departure concentrations. |
| | Krewski D; Westphal M; Paoli G; Croteau M; Al-Zoughool M; Andersen M; Chiu W; Cote I. (in preparation). A framework for the next generation of risk science. |
| knowledgebases | Provide an alternative approach for storing and searching the complete networks of highly interconnected information produced by linking bioassays and pathways. Developed decades ago to codify human knowledge so that they could be used to efficiently support decisions, knowledgebases are finding practical applications in meaningfully organizing vast amounts of linked biological data using ontologies. |
| Kyoto Encyclopedia of Genes and Genomes (KEGG) | A database resource that integrates genomic, chemical, and systemic functional information. In particular, gene catalogs from completely sequenced genomes are linked to higher level systemic functions of the cell, the organism, and the ecosystem. KEGG is a reference knowledgebase for integration and interpretation of large-scale data sets generated by genome sequencing and other high-throughput experimental technologies. |
| | Kanehisa Laboratories. (2012). KEGG: Kyoto encyclopedia of genes and genomes. Available online at http://www.genome.jp/kegg/ (accessed February 22, 2013). |
| lift | Lift is a measure of how much better prediction results are using a model than could be obtained by chance. For example, say 2% of customers who receive a catalog in the mail make a purchase, and when a model is used to select catalog recipients, 10% make a purchase. The lift for the model would be 10/2 or 5. |
| | Oracle. (2013). Glossary: "Lift". Available online at http://docs.oracle.com/cd/B28359_01/datamine.111/b28129/glossary.htm (accessed March 20, 2013). |

*This document is a draft for review purposes only and does not constitute Agency policy. Do not cite or quote.*

September 2013                                                    B-7

| Glossary Term | Description |
|---|---|
| **Meta Data Viewer** | A publicly available graphical display software program that can be used to graph animal and human data. Meta Data Viewer can display up to 15 text columns and to graph 1–5 numerical values. Users can sort, group, and filter data and examine patterns of findings across studies. Users can use the program and any associated National Toxicology Program (NTP) data files for their own purposes, including for use in publications. |
| | National Toxicology Program (NTP). (2012). Meta Data Viewer. Available online at http://ntp.niehs.nih.gov/?objectid=1DF7D40E-A957-9727-733C9B89E243634B (accessed September 27, 2012). |
| **microarray analysis** | The simultaneous analysis, on a microchip, of multiple samples or targets arranged in an array format. |
| | National Center for Biotechnology Information (NCBI). (2012). Microarray Analysis. Available online at http://www.ncbi.nlm.nih.gov/mesh/?term=microarray%20analysis (accessed September 27, 2012). |
| **microarray technology** | A developing technology used to study the expression of many genes at once. It involves placing thousands of gene sequences in known locations on a glass slide called a gene chip. A sample containing DNA or RNA is placed in contact with the gene chip. Complementary base pairing between the sample and the gene sequences on the chip produces light that is measured. Areas on the chip producing light identify genes that are expressed in the sample. |
| | National Human Genome Research Institute (NHGRI). (2012). Talking Glossary of Genetic Terms. Available online at http://www.genome.gov/glossary/index.cfm?id=125&textonly=true (accessed September 27, 2012). |
| **mode of action** | The key steps in the toxic response after chemical interaction at the target site that is responsible for the physiological outcome or pathology of the chemical; how chemicals perturb normal biological function. |
| | U.S. Environmental Protection Agency (EPA). (2012). Glossary of Terms: Methods of Toxicity Testing and Risk Assessment. Available online at http://www.epa.gov/opp00001/science/comptox-glossary.html (accessed April 2, 2013). |
| **mode-of-action-based *in vitro* toxicity pathway assays** | Fit-for-purpose assays using human cells to assess biological pathway perturbations based on specific or generic modes of action. The suite of these assays would form the test battery for safety assessment. |
| | Krewski D; Westphal M; Paoli G; Croteau M; Al-Zoughool M; Andersen M; Chiu W; Cote I. (in preparation). A framework for the next generation of risk science. |

*This document is a draft for review purposes only and does not constitute Agency policy. Do not cite or quote.*

September 2013                                                    B-8

| Glossary Term | Description |
|---|---|
| molecular epidemiology | Referring to the application of molecular biology to answer epidemiological questions. The examination of patterns of changes in DNA to implicate particular carcinogens and the use of molecular markers to predict which individuals are at highest risk for a disease are common examples. Molecular epidemiology incorporates molecular markers of exposure and biological change into population-based studies; integrates knowledge of the human genome into epidemiological studies to understand genetic susceptibility and gene-environment interaction in disease causation.<br><br>National Center for Biotechnology Information (NCBI). (2012). Molecular Epidemiology. Available online at http://www.ncbi.nlm.nih.gov/mesh?term=molecular%20epidemiology (accessed September 27, 2012); Krewski D; Westphal M; Paoli G; Croteau M; Al-Zoughool M; Andersen M; Chiu W; Cote I. (in preparation). A framework for the next generation of risk science. |
| omics | Refers to a broad field of study in biology, ending in the suffix "-omics" such as genomics, proteomics, transcriptomics.<br><br>U.S. Environmental Protection Agency (EPA). (2012). Glossary of Terms: Methods of Toxicity Testing and Risk Assessment. Available online at http://www.epa.gov/opp00001/science/comptox-glossary.html (accessed April 2, 2013). |
| ontology | Defines types of data (e.g., chemicals, genes, assays, interactions, pathways, cells, species) and their interrelationships (chemicals "activate" proteins; assays "measure" changes in proteins; genes are "part of" pathways, etc.). |
| phenotype | An individual's observable traits, such as height, eye color, and blood type. The genetic contribution to the phenotype is called the genotype. Some traits are largely determined by the genotype, while other traits are largely determined by environmental factors.<br><br>National Human Genome Research Institute (NHGRI). (2012). Talking Glossary of Genetic Terms. Available online at http://www.genome.gov/glossary/index.cfm?id=152&textonly=true (accessed September 27, 2012). |
| polymerase chain reaction (PCR) | A method for amplifying a DNA base sequence using a heat-stable polymerase and two 20-base primers, one complementary to the (+) strand at one end of the sequence to be amplified and one complementary to the (-) strand at the other end. Because the newly synthesized DNA strands can subsequently serve as additional templates for the same primer sequences, successive rounds of primer annealing, strand elongation, and dissociation produce rapid and highly specific amplification of the desired sequence. PCR also can be used to detect the existence of the defined sequence in a DNA sample.<br><br>Department of Energy (DOE). (2010). Human Genome Project Information: Genome Glossary. Available online at http://www.ornl.gov/sci/techresources/Human_Genome/glossary/glossary_p.shtml (accessed September 27, 2012). |

*This document is a draft for review purposes only and does not constitute Agency policy. Do not cite or quote.*

September 2013           B-9

| Glossary Term | Description |
|---|---|
| **principal components analysis (PCA)** | A mathematical procedure that transforms several possibly correlated variables into a smaller number of uncorrelated variables called principal components. |
| | National Center for Biotechnology Information (NCBI). (2012). Principal Components Analysis. Available online at http://www.ncbi.nlm.nih.gov/mesh?term=principal%20component%20analysis (accessed September 27, 2012). |
| **probe** | Single-stranded DNA or RNA molecules of specific base sequence, labeled either radioactively or immunologically, that are used to detect the complementary base sequence by hybridization. |
| | Department of Energy (DOE). (2010). Human Genome Project Information: Genome Glossary. Available online at http://www.ornl.gov/sci/techresources/Human_Genome/glossary/glossary_p.shtml (accessed September 27, 2012). |
| **proteomics** | The study of the function of all expressed proteins. |
| | U.S. Environmental Protection Agency (EPA). (2012). Glossary of Terms: Methods of Toxicity Testing and Risk Assessment. Available online at http://www.epa.gov/opp00001/science/comptox-glossary.html (accessed April 2, 2013). |
| **quantitative structure activity relationship (QSAR)** | A mathematical relationship between a quantifiable aspect of chemical structure and a chemical property or reactivity or a well-defined biological activity, such as toxicity. Using a sample set of chemicals, a relationship is established between one or many physical-chemical properties a chemical possesses due to its structure and a chemical property or biological activity of concern. This mathematical expression is then used to predict the chemical property or biological response expected from other chemicals with similar structures. It is based on the presumption that similar molecules or chemical structures have similar properties or biological activities or toxicity potential. |
| | U.S. Environmental Protection Agency (EPA). (2012). Glossary of Terms: Methods of Toxicity Testing and Risk Assessment. Available online at http://www.epa.gov/opp00001/science/comptox-glossary.html (accessed April 2, 2013). |
| **QSAR Toolbox** | A software application intended for use by government, the chemical industry, and other stakeholders in filling gaps in (eco)toxicity data needed for assessing the hazards of chemicals. The Toolbox incorporates information and tools from various sources into a logical workflow. Crucial to this workflow is grouping chemicals into chemical categories. The seminal features of the Toolbox are identification of relevant structural characteristics and the potential mechanism or mode of action of a target chemical, identification of other chemicals that have the same structural characteristics or mechanism/mode of action (or both), and use of existing experimental data to fill the data gap(s). |
| | QSAR Toolbox. (2012). About: What does the QSAR Toolbox do? Available online at http://www.qsartoolbox.org/ (accessed September 28, 2012). |

*This document is a draft for review purposes only and does not constitute Agency policy. Do not cite or quote.*

September 2013                                    B-10

| Glossary Term | Description |
| --- | --- |
| reverse toxicokinetics (RTK) | Also known as reverse dosimetry, refers to the use of a pharmacokinetic model to estimate external dose (exposure) from a known internal concentration. The method uses a one-compartment model and makes default assumptions such as chemicals are eliminated wholly through metabolism and renal excretion; renal excretion is a function of the glomerular filtration rate and the fraction of unbound chemical in the blood (i.e., no active transport); and oral absorption is 100%. Using these assumptions, the plasma concentration of the chemical at steady state per unit dose then can be estimated. The two experimental chemical-specific parameters required to generate an estimate are the rate of disappearance of parent via hepatic metabolism (intrinsic clearance) and fraction bound (or conversely unbound) to plasma proteins. Both parameters can be measured experimentally in a relatively high-throughput manner.

Judson RS; Kavlock RJ; Setzer RW; Hubal EA; Martin MT; Knudsen TB; Houck KA; Thomas RS; Wetmore BA; Dix DJ. (2011). Estimating toxicity-related biological pathway altering doses for high-throughput chemical risk assessment. Chem Res Toxicol Chem Res Toxicol 24 (4): 451-462. http://dx.doi.org/10.1021/tx100428e |
| rule | A rule describes an association between elements on the left-hand side of the rule and items on the right-hand side of the rule. For instance, the rule [diapers, cola] => [milk] in a supermarket database might mean that when customers bought diapers and cola, they also purchased milk. |
| ruleset | A ruleset is a collection of one or more rules that can be associated with a realm authorization, factor assignment, command rule, or secure application role. The ruleset will be "true" or "false" based on evaluation of each rule in the ruleset and the evaluation type for the ruleset, which can be "all true" or "any true."

Oracle. (2013). 5 Configuring Rule Sets. Available online at http://docs.oracle.com/cd/B28359_01/server.111/b31222/cfrulset.htm#DVADM70150 (accessed March 20, 2013). |
| SNPs | Refers to single nucleotide polymorphisms, which are single nucleotide variations in a genetic sequence that occur at appreciable frequency in the population.

National Center for Biotechnology Information (NCBI). (2012). SNPs. Available online at http://www.ncbi.nlm.nih.gov/mesh?term=SNPS (accessed September 28, 2012). |
| stem cell biology | A branch of biology that studies and develops stem cells, which are cells with the ability to divide for indefinite periods in culture and to give rise to specialized cells.

The National Institutes of Health (NIH). (2009). Stem Cell Basics. Available online at http://irp.nih.gov/catalyst/v19i6/systems-biology-as-defined-by-nih (accessed September 28, 2012). |

*This document is a draft for review purposes only and does not constitute Agency policy. Do not cite or quote.*

September 2013        B-11

| Glossary Term | Description |
|---|---|
| systems biology | A scientific approach that combines the principles of engineering, mathematics, physics, and computer science with extensive experimental data to develop a quantitative as well as a deep conceptual understanding of biological phenomena, permitting prediction and accurate simulation of complex (emergent) biological behaviors. |
| | Wanjek, C. (2011). Systems biology as defined by NIH. The NIH Catalyst 19 (6): November-December. http://irp.nih.gov/catalyst/v19i6/systems-biology-as-defined-by-nih. |
| TOM (topological overlap matrix) heat map | A graphical representation in which the rows and columns represent genes in a symmetric manner; the color intensity represents the interaction strength between genes. |
| | Wang I.; Zhang B; Yang X; Stepaniants S; Zhang C; Meng Q; Peters M; He Y; Ni C; Slipetz D; Crackower MA; Houshyar H; Tan CM; Asante-Appiah E; O'Neill G; Luo MJ; Theiringer R; Yuan J; Chiu C; Lum PY; Lamb J; Boie Y; Wilkinson HA; Schadt E; Dai H; Roberts C. (2012). Systems analysis of eleven rodent disease models reveals an inflammatome signature and key drivers. Molecular Systems Biology 8 594. |
| toxicity pathways | The 2007 NRC report on Toxicity Testing in the 21st Century envisioned that new technologies will help us better understand how chemicals perturb normal biological function, and thus identify toxicity pathways. Potential toxic effects of chemicals would be predicted based on *in vitro* bioactivity profiles derived from a chemical's effects on cellular molecules and processes. The interpretation of chemically induced perturbations in toxicity pathways depends on linking *in vitro* effects with adverse outcomes *in vivo*, and on computer modeling that extrapolates to predicted responses in whole tissues, organisms, and populations based on realistic human or environmental exposures. |
| | U.S. Environmental Protection Agency (EPA). (2012). Glossary of Terms: Methods of Toxicity Testing and Risk Assessment. Available online at http://www.epa.gov/opp00001/science/comptox-glossary.html (accessed April 2, 2013). |
| toxicogenomics | Study of the roles that genes play in the biological responses to environmental toxicants and stressors by the collection, interpretation, and storage of information about gene and protein activity. |
| | U.S. Environmental Protection Agency (EPA). (2012). Glossary of Terms: Methods of Toxicity Testing and Risk Assessment. Available online at http://www.epa.gov/opp00001/science/comptox-glossary.html (accessed April 2, 2013). |
| transcription | The biosynthesis of RNA carried out on a template of DNA. The biosynthesis of DNA from an RNA template is called reverse transcription. |
| | National Center for Biotechnology Information (NCBI). (2012). Transcription. Available online at http://www.ncbi.nlm.nih.gov/mesh/68014158 (accessed September 27, 2012). |

*This document is a draft for review purposes only and does not constitute Agency policy. Do not cite or quote.*

September 2013
B-12

| Glossary Term | Description |
|---|---|
| transcriptome | The pattern of gene expression, at the level of genetic transcription, in a specific organism or under specific circumstances in specific cells.<br><br>National Center for Biotechnology Information (NCBI). (2012). Transcriptome. Available online at http://www.ncbi.nlm.nih.gov/mesh/68059467 (accessed September 27, 2012). |
| transcriptomics | The study of gene expression at the RNA level.<br><br>U.S. Environmental Protection Agency (EPA). (2012). Glossary of Terms: Methods of Toxicity Testing and Risk Assessment. Available online at http://www.epa.gov/opp00001/science/comptox-glossary.html (accessed April 2, 2013). |
| transgenic | Produced from a genetically manipulated egg or embryo; containing genes from another species.<br><br>National Center for Biotechnology Information (NCBI). (2012). Transgenic. Available online at http://www.ncbi.nlm.nih.gov/mesh/?term=transgenic (accessed September 27, 2012). |
| translation | The process of translating the sequence of a messenger RNA (mRNA) molecule to a sequence of amino acids during protein synthesis. The genetic code describes the relationship between the sequence of base pairs in a gene and the corresponding amino acid sequence that it encodes. In the cell cytoplasm, the ribosome reads the sequence of the mRNA in groups of three bases to assemble the protein.<br><br>National Human Genome Research Institute (NHGRI). 2012. Talking Glossary of Genetic Terms. Available online at http://www.genome.gov/glossary/index.cfm?id=200&textonly=true (accessed September 28, 2012). |
| translesion synthesis | A mechanism for DNA damage tolerance that allows the DNA replication machinery to move beyond a DNA lesion or abasic site (i.e., a site that lacks a DNA base). |
| Virtual Tissue (v-Tissues™) Models | In silico cross-scale models of cellular organization and emergent functions used to better understand disease progression. Tissues are the clinically relevant level for diagnosing and treating the transition from normal to adverse states in chemical-induced toxicities leading to cancer, immune dysfunction, developmental defects, and more. Currently, in vivo rodent experiments are used to evaluate tissue-level effects of altered molecular and cellular function; however, the extrapolation of animal models to humans is often uncertain. v-Tissues™ aim to simulate key molecular and cellular processes computationally in the context of normal tissue biology to: (1) help understand complex physiological relationships, and (2) predict adverse effects due to chemicals. As the number of chemicals in consumer products, the workplace, and the environment continues to rise, v-Tissues™ offers the promise of a more efficient, effective, and humane approach for evaluating their impact on human health.<br><br>U.S. Environmental Protection Agency (EPA), Computational Toxicology Research Program. What are Virtual Tissues? (2012). Available online at http://www.epa.gov/ncct/virtual_tissues/what.html (accessed September 27, 2012). |